Greening Africana Studies

Greening Africana Studies

Linking Environmental Studies with Transforming Black Experiences

Rubin Patterson

TEMPLE UNIVERSITY PRESS
Philadelphia Rome Tokyo

TEMPLE UNIVERSITY PRESS
Philadelphia, Pennsylvania 19122
www.temple.edu/tempress

Published 2015

Library of Congress Cataloging-in-Publication Data

Patterson, Rubin.
Greening Africana studies : linking environmental studies with transforming black experiences / Rubin Patterson.
pages cm
Includes bibliographical references and index.
ISBN 978-1-4399-0871-6 (hardback : alk. paper) —
ISBN 978-1-4399-0872-3 (paper : alk. paper) —
ISBN 978-1-4399-0873-0 (e-book) 1. Sustainable development—Africa. 2. African American neighborhoods—Environmental aspects. 3. Environmental degradation. 4. Climate change. I. Title.
HC800.Z9E5593 2015
338.96′07—dc23

2014018294

∞ The paper used in this publication meets the requirements of the American National Standard for Information Sciences—Permanence of Paper for Printed Library Materials, ANSI Z39.48-1992

Printed in the United States of America

9 8 7 6 5 4 3 2 1

Contents

Acknowledgments

I am grateful for the contributions of many people who have in various ways helped me complete this book. First and foremost, I thank my dear friend and fellow Howard University graduate, Cynthia Hewitt, who is a professor of sociology at Morehouse College. She is the person who managed to convince me to think critically, for the first time, about our environment and the need for humans to engage with it sustainably. During the 2005–2006 academic year, I took a sabbatical from my former campus, the University of Toledo (UT), to be a visiting professor at Morehouse so that Cynthia and I could work on projects involving African brain circulation, which is a focused and intentional form of African transnationalism. In the course of working on those projects, she and I engaged in many debates about the environment, all of which I lost. My losing those deliberations is partly explained by Cynthia's formidable debating skills, but some of it, is also simply because there is little to debate about the fact that human activities are breaching the biophysical limits of the planet to our peril. And as injustice would have it, the people whose activities are contributing the least to the breaching and polluting are unjustly experiencing the most intense effects of climate change and other transformations of the environment.

After completing my sabbatical and returning to UT in 2006, I continued to read voraciously on the broad subject of the environment as if I were satisfyingly scratching an itch whose persistence I welcomed. Almost immediately, the environment had become my principal research agenda. Most of my scholarly publications and my professional activities since my yearlong intellectual sparring with Cynthia have had to do with some aspect of the environment: for example, I created the first Africana studies program in the nation that focused on sustainability; I created a clutch of new sustainability courses; I recruited professors to teach sustainability courses in the disciplines of Africana studies and sociology in the United States and in Africa; for years I wrote a monthly green column in the African American Toledo newspaper the *Sojourner's Truth* to instigate critical thinking about the environment and the black experience; and I engaged in sustainability projects in Toledo and in Southern Africa. All of these accomplishments stemmed from my series of debates about the environment to Cynthia. I have never gained so much from losses!

One of my many green publications is an article I wrote with my former graduate student, Nicole Lambert, who is now completing her doctoral degree in sociology at the University of Colorado at Boulder. Nikkie and I published that piece, which included content on greening Africana studies, in the *Journal of Black Studies* in 2008. I am as grateful to her for allowing me to use materials from the article as I am to her for the data compilation she completed for it. And then there is Beatrice Miringu, an environmental scientist for the city of Toledo's Department of Environmental Services. Beatrice helped me sort out issues related to the two types of locally unwanted land uses (LULUs) covered in this book—brownfields and toxics release inventory (TRI) facilities. Regarding brownfields, there seems to be no way of knowing precisely the number of sites in existence because of different interpretations of the term in the thousands of jurisdictions around the country and variations in the willingness of local governments to acknowledge the existence of some brownfields in their respective jurisdictions.

For helping me imbue the book with a sense of history, I thank my friend and former UT colleague Professor Angela M. Siner. Instead of a book that merely presents accurate historical facts and a narration about postulated cause and effect, Angela helped me produce a book that captures the flow of the historical development of environment issues in the context of the lived black experience.

I also must acknowledge the invaluable green intellectual support I received from two other former UT colleagues, Professors Seamus Metress and Eileen Metress. For decades they have provided Northwest Ohio with a principled and critical voice for environmental sustainability that resonated with me. I tried to capture, as a current that runs through this book, their sincerity and tireless effort to advance knowledge and activism for a sustainable environment.

A final former colleague to thank is Charlene Gilbert, who was the chair of women's and gender studies at the University of Toledo. The few bucks I spent one afternoon for coffee and conversation with her probably yielded the most lucrative return that I could ever receive from such a small investment. Charlene helped me write more confidently on the topics of black feminism and Africana womanism, which are covered in a section of Chapter 2. After comprehending and synthesizing the education she provided me, I became more self-assured in presenting my ideas on those subjects. That said, I am solely responsible for any inaccuracies that may be found in that section of Chapter 2.

To the students whose help I was fortunate to receive, I am grateful. First, Giselle Thompson, who was my graduate assistant, helped me prepare the full manuscript for submission to Temple University Press. On occasion we worked in the office in the small hours of the morning to meet deadlines. Giselle is now a doctoral student at York University in Toronto. I also acknowledge my doctoral graduate assistant, Edwina Kofi-Opata, and my research assistant, Nomathemba B. Sibanda, who helped in proofreading the final versions of the book. It would have been difficult, if not impossible, to complete the final review of the copyedited manuscript if it were not for Nomathemba. And then there were Katricia Welliford and Steven Jackson (a.k.a. Kumasi), who helped with the laborious, tedious work of identifying brownfields in the hundreds of U.S. cities where Africana studies programs exist. The final two students whom I must recognize and thank are Chelsi Holder and Laura Hoffacker. Because both are quite skilled at data mining and generating illustrations, they were of tremendous help to me.

I also thank Glenda Sneed for her quiet and indispensable contributions, which made this book possible.

Last, but in no way least, I thank Dr. Forrestine Barnes for her editorial help with the book. I was fortunate during my final semester at Howard University when my dissertation manuscript was assigned to Dr. Barnes for reviewing. After working with her through the editing process, I felt that I had learned as much about being an effective writer as I had about being an effective researcher of my topic. Somehow I had the wisdom to entreat her, over the past two decades, to continue assisting me, especially in editing my manuscripts for publication and serving as a copyeditor of the peer-reviewed journal (*Perspectives on Global Development and Technology*) for which I was the chief editor. It is fitting that I have returned to Howard University, where I was first introduced to Dr. Barnes, from whom I have greatly benefited as a writer.

Greening Africana Studies

Introduction

Bridging Africana Studies and Environmental Studies

Contextualizing the Study

Budding intellectuals who are attracted to Africana studies and environmental studies are, of course, curious about the subject matter of these half-century-old academic fields, but they also want to know that their studies prepare them to become effective change agents, empowered to bring about and administer more equitable and sustainable societies. This book responds to the inquiries and needs of such individuals. First, however, we must recognize the two fields of study as complementary in their respective missions. For this reason, the primary aim of this book is to explore and rectify the disconnect between Africana studies and environmental studies within the academy—a disconnect that exists despite their similarities on multiple levels and the numerous ways the two can support one another's scholarly and activist interests.

In addition to examining some of the key similarities of Africana and environmental studies, this book discusses why and how the two should be integrated. To this end, it provides comprehensive knowledge of the two fields, focusing on the mission and major paradigms that identify their respective curricula, research interests, and practices. It also shows that several of the major paradigms that underlie academic activity in both Africana and environmental studies compete for dominance in each field and that there are opportunities to collaborate across them.

On the surface, the most conspicuous similarity between Africana studies and environmental studies is the diversity of their appellations or naming conventions. For instance, on one side we have Africana studies, African American studies, Afro-American studies, Pan-African studies, black studies, and Africology. Africana studies looks at the black experience in America primarily and in the rest of the diaspora and Africa secondarily. All the aforementioned appellations to some extent cover African American studies; African studies, however, does not necessarily cover African America. African studies examines the historical and contemporary social dynamics of lived experiences on the African continent primarily and may include black America and U.S. politico-economic and cultural forces vis-à-vis Africa secondarily. In other words, Africana studies primarily covers African America, while African studies mainly covers the African continent. The study of the environment encompasses environmental studies, environmental justice, and sustainability studies. Environmental studies could be considered the broad canvas on which environmental justice and sustainability studies are painted. After all, environmental studies examines human-environment interactions in general, whereas environmental justice and sustainability studies focus on the ways humans engage with the environment and share the goals of just distribution of resources, regulations on environmental degradation, and sustainable outcomes. A related field, environmental sciences, focuses primarily on the physics, chemistry, and biology of the environment and only secondarily on its so-called softer side—policy, aesthetics, culture, and economics.

The point is that both Africana studies and environmental studies, unlike traditional disciplines, have multiple names for single intellectual endeavors. Some early critics of these fields of study suggested that they were not ready for academic prime time and that they were not deserving of scarce university resources. Some of the harshest criticisms identified them as little more than products of political largess. Traditional disciplines such as sociology, biology, history, economics, and physics, so the argument goes, settled on the parameters, bedrock assumptions, key questions, competing paradigms, and, perhaps most elemental, consistent naming before taking the national academic stage. The inability of Africana studies and environmental studies to reach uniformity on a name for their departments and programs undermines their credibility in the eyes of critical observers. However, the frequency of these criticisms is diminishing as the innovative contributions of these fields to the academy over the past half-century have been recognized.

Referring to Africana studies, Martha Biondi notes, "Many critics, both internal and external to Black studies, criticized it on two interrelated grounds: they claimed that it lacked curricular coherence and that by not having a single methodology it failed to meet the definition of a discipline. As a result, many educators in the early Black studies movement pursued a two-pronged quest: a standardized curriculum and an original, authoritative methodology" (2012: 241). Africana studies scholars such as Abdul Alkalimat have long lamented the lack of a uniform curriculum and pedagogy that would define the field. Conversely, Rhett Jones, a cofounder of Africana studies at Brown University, has been a longtime critic of the uniformity approach. Darlene Clark Hine hints that if there is any silver lining to the absence of curriculum standardization, it is what it suggests about the ever-changing and evolving nature of Africana studies: "The rapid proliferation of knowledge in the field is a strong argument in support of institutional flexibility" (1997: 11). While Alkalimat and Biondi lament the lack of uniformity, Hine and Jones seem to argue for innovational latitude brought on by the absence of standardization. Both sides of the argument have merit, but Africana studies, which builds on the rich diversity of the African and American contexts, may need to legitimize its existence in academia like most previously established fields. It can achieve such legitimacy only through the creation of specific approaches unique to Africana studies that combine structured and unstructured components and allow for intellectual creativity.

The notion of knowledge for knowledge's sake has never taken root in either Africana or environmental studies. Both have an academic mission and an unapologetic social mission. The phrase "academic excellence and social responsibility," coined by Abdul Alkalimat when Africana studies was undergoing its initial growth spurt, is actually applicable to environmental studies also. Both fields were founded in political struggle, in efforts to better understand and address the ills of racism, social inequality, and environmental degradation and to empower blacks and other subalterns as well as to advance environmental sustainability. Fighting against social ills and fighting for empowerment and sustainability are political tasks for which Africana and environmental studies are totally unapologetic and well suited, as they are for the task of advancing innovative and rigorous scholarship.

Africana studies documents and challenges racism while promoting the agency of people of African descent in responding to and reshaping their social settings. Its well-known mission is to improve the life experiences of black people as well as to improve sources of knowledge about those experiences. Students of Africana studies identify a specific time period (e.g., 2010s) and a given issue (e.g., redeveloping brownfield sites in black neighborhoods with renewable energy production or urban agriculture) and then examine the use of one or some combination of the field's paradigms. In other words, for this issue in this time frame—as is the case for any black experience issue in a given time frame—the student will ask: "What are the specific institutional ways in which racism and other structural factors have limited the options of brownfield remediation in black neighborhoods?" "Why are the brownfields so disproportionate in black neighborhoods in the first place?" "How have African Americans responded productively and unproductively to the racial imbalances that militate against the success of brownfield redevelopment?" Responses to these questions lead to discussions about African Americans who have taken proactive measures against such racist practices, not simply responding and adjusting to them but trying to reshape social institutions in a way that would lead to more brownfield redevelopment and more renewable energy. In essence, Africana studies, is the scientific study of and humanistic reflection on holistic aspects of contemporary and historical black thought and practice.

Similarly, environmental studies documents and challenges environmental degradation while documenting and promoting the agency of environmentally conscious and politically engaged individuals and institutions in reshaping human-environment interactions. Students in the field engage primarily environmental issues that concern us today, but they also engage such issues historically and archaeologically.

Africana studies is oriented toward better understanding and significant improvement of the black experience, and environmental studies is oriented toward a better understanding of negative human impacts on the environment and ways to reduce them. The two fields are progressive in that, if they are to initiate means for improving the black experience and for promoting more sustainable human-environment interactions, they must question and critique the existence, activities, and culture of institutions and individuals in society as each impacts blacks and the environment.

By definition and objective, Africana and environmental studies have oppositional stances and identities when compared to economics and business, with their inherent focus on finance and marketing—the principal disciplinary pillars of the establishment. Like Africana and environmental studies, economics is a liberal arts discipline, but it is not nearly as progressive in the sense that, at least in the United States, it is essentially restricted from questioning the fundamental tenets of capitalism. It is limited to identifying ways of making capitalist economies work more efficiently and to reducing the duration and severity of the recessions that are fixtures in capitalist economies. Likewise, business, by nature, does not question or debate the fundamental merits and demerits of capitalist enterprises. Africana and environmental studies, however, routinely examine propositions concerning the inevitability of labor exploitation, social inequality, and environmental degradation in all, including capitalist, economies.

Perry Hall argues that Africana studies disrupts Western narratives of development "by decentering the hegemonic Eurocentric approaches to knowledge production" (2010: 31). Additionally, Africana studies disrupts the "Eurocentric tendency to marginalize and pathologize blackness" while centralizing and celebrating whiteness (22). Analysis of people of African descent as "the other" facilitates a narrative of the social pathologies of poverty, failures in education, and crime-ridden neighborhoods as a direct result of "the other's" inherent deficiencies—culturally or genetically induced. Conversely, Africana studies makes black people the central agents rather than reducing them to "the other." Beyond merely documenting and describing social ills in the black community, as was long the case with traditional disciplines, Africana studies examines and reflects on black people's cultural, economic, and political responses to institutions that maintain racial inequality. In addition, it centralizes black agency in positive experiences for black people and in reshaping visions of the black presence in our society, culturally with its music, style, and attitude and politico-economically with its overwhelming preference for politicians and policies that seek to end economic and racial inequality.

Environmental studies can also disrupt and decenter the hegemonic Eurocentric approach to economic production and human-environment interactions. For it to produce the knowledge and activities that will move society to a sustainable economy, the transformation of our energy sources from fossil fuels to renewables is absolutely paramount. Such an objective is most challenging and most disruptive, as we live in an era in which our food, our clothing, and virtually everything else depend on fossil fuels, and as a consequence we have made the very people we need to defeat politically to achieve sustainability more powerful by enriching them with wealth from meeting our ever-growing energy needs. And that enrichment is used in part to purchase political support from elected and appointed officials and to confuse the public with disinformation such as the denial of climate change. Among the world's five hundred largest companies in 2011, four of the top ten were Western oil companies, two were Chinese oil companies, and another was a Chinese power company. The eighth top-ten company, Toyota, depends completely on oil. What this suggests is that fossil fuel industries can purchase political power as a means to dominate much of the law and policy making in Washington.

Alas, one area in which environmental studies has not disrupted and decentered Eurocentric hegemonic thinking is race. The field engages in color-blind discourse, thereby supposedly obviating the need for, if not delegitimizing, critics of racialization. Bringing racialization to the table could prove to be one of the greatest contributions of Africana studies to environmental studies.

Africana and environmental studies have contributed to the decentering of Eurocentric knowledge building and economic production. Because of Africana studies, traditional academic disciplines have begun to cover transformative black experiences as a centralized rather than as a pathologized narrative of "the other's" social ills. Africana courses at elite, predominately white universities—Oberlin College and the University of Texas among others—are easily among the most popular on campus. Students who have been steeped in Africana studies have not only carried on with such myth debunking but have also gone on to create new knowledge and engage civically based in part on this exposure. Furthermore, half of all Africana studies programs are at major research universities, and another 14 percent are at elite liberal arts colleges (Rojas 2007). As Gwendolyn Etter-Lewis (2006) notes, discourses do not just reflect reality; they help re-create it. So, as Africana studies has played a role in changing discourses, it has thereby also played a role in changing realty.

Similarly, as a result of environmental studies on campuses and in the broader environmental movement, industrial polluters now feel compelled to show their green bona fides, which are often contrived, in order to maintain and grow market share. It is not insignificant that this field has been among the fastest growing in the academy in the twenty-first century. Indeed, according to the Association for the Advancement of Sustainability in Higher Education (2009), over a hundred new degree programs focusing on energy and sustainability were created in 2009 alone. Forty years ago, Africana studies experienced similar stunning growth.

Both Africana and environmental studies gained academic status principally because of social movements beyond campus. For Africana studies, it was the civil rights and black power movements, whereas for environmental studies it was heightened green consciousness among the public as well as the growing environmental movement in all of its permutations in the wake of Rachel Carson's *Silent Spring* (1964). As Paul Hawken (2007) notes, the environmental movement is the largest social movement ever to have existed. Moreover, unlike with the civil rights and black power movements, there is no leader for hegemonic forces to try to eliminate and thereby upend the movement—it is too lateral and diffuse to be stopped.

Since Africana and environmental studies are not establishment disciplines in the mold of economics and business, they will continue to rely in part on social movement support for their existence.

Environmental studies, unlike Africana studies, will continue on a faster growth path because of the continued energetic growth of the environmental movement as well as many new opportunities in the nonprofit and private sectors. Africana studies is currently without such systematic extra-academic forces in the community or in the broader society, which will likely limit its further growth outside of the elite research and liberal arts institutions where significant investment in the field continues.

Africana and Environmental Studies: They Need Each Other to Fulfill Their Missions

The notable successes in Africana and environmental studies should be kept in perspective, considering how much more work needs to be done in the areas of racial inequality and environmental degradation and its attendant climate chaos. This work is at the heart of why it is imperative that the two fields work collaboratively. A clearly recognized fact, however, is that the academic and social mission of neither can be successfully realized unless the political power of fossil fuel is greatly diminished and ultimately eclipsed by renewable energy and other sustainable industries. That is perhaps patently obvious with environmental studies and less so with Africana studies. The former's mission of creating a sustainable economy and environment cannot be achieved as long as fossil fuel industries wield substantial influence in Washington.

The larger fossil fuel industry constitutes an interest group, whose products power over three-quarters of human activity, that uses political parties as one instrument to meet its objectives, one of which is to retain for as long as possible its status as the subsidized energy source of choice and to prevent renewable energy sources from reaching grid parity or becoming competitive. The industry finds it politically expedient to dominate the Republican Party so that these outcomes can be achieved. As an example, consider the fact that there were thirty-seven Senate seats open in the 2010 electoral season; among the forty-eight Republican candidates for those seats, only one agreed that human-made, or anthropogenic, global warming was real and that fossil fuels were a major factor in its development. Former representative

Mike Castle of Delaware, the lone exception, would have won easily against a Democratic opponent, but the fossil fuel–backed Tea Party and Americans for Prosperity saw to it that he was defeated in the primary. One reading of this situation is that the fossil fuel industry was willing to lose one senate seat as a way to warn Republican candidates that they must deny climate change and support fossil fuels while only rhetorically—at best—supporting renewables.

In other words, it is nearly impossible for Republican candidates to win primaries for statewide, congressional, or presidential elections if there is any suggestion that they advocate any of the following: significantly taxing the pollution-emitting carbon that contributes to climate change, removing fossil fuel subsidies, significantly subsidizing renewable energy to gain grid parity with coal and natural gas, and further restricting toxic chemicals.

Mitt Romney is a telling example of the fossil fuels industries' agenda. As governor of liberal Massachusetts, he pushed to close old cold-fired power plants and supported wind and solar energy. Nevertheless, as a Republican presidential candidate, "he [was] a proclaimed skeptic on global warming, a champion of oil and other fossil fuels, a critic of federal efforts to develop cleaner energy sources and a sworn enemy of the Environmental Protection Agency" ("Energy Etch a Sketch" 2012). Political parties are well known as instruments of power for coordinated and influential interest groups. The Republican Party is one instrument of choice for the fossil fuel industry. This sounds partisan, but even a cursory look at fossil fuel interests' political contributions reveals that they are heavily skewed toward Republicans and other deregulatory politicians.

No one should be surprised that discussions of environmental studies are inseparable from discussions of politics. The same conclusion applies to Africana studies, which presupposes politics given that it was founded in political struggles. Additionally, it is no secret that Republicans have won most of the white vote and have dominated southern electoral votes since Lyndon Johnson signed the Civil Rights Act of 1964. For all presidential contests between 1992 and 2008, between 59 and 72 percent of the Republican nominee's electoral votes came from the South. The South is crucial to the fossil fuel industry's agenda, but it remains the industry's primary province because of the stoking and exploitation of latent racial animus (Patterson 2013b). This issue is, of course, of special concern in Africana and environmental studies programs.

In the United States, as a rule, racial fears and anxieties tend to be higher in areas where blacks constitute a large percentage of the population. In the Pacific Northwest and the northern New England states, for instance, where black populations are extremely small, discordant race relations are few. Conversely, in the Deep Peripheral South, where black populations range from 12 to 37 percent of states' populations, racial discord is consequential. Whites who reside in communities that include or are adjacent to large black populations are more racially conservative than are those who reside in areas with smaller black populations in the same states (Glaser 1994; Valentino and Sears 2005).

As Oliver Cox (1948) and other scholars have long noted, elites manipulate a heightened sense of economic threat among white workers based on race, not class, that springs from intense zero-sum thinking played out as racial strife. And when such thinking overlies the belief that a large black population threatens white workers—added to the fact that the South was home to the most brutal forms of chattel slavery and the cruelest and longest forms of Jim Crow, as well as the fact that the region has long been the poorest and worst educated in the country—it is understandable why the fossil fuel industry focuses on the South, where racial fissures can be stoked more easily and fruitfully for their global private gain.

Given that Africana studies programs (those focusing on U.S. issues) primarily study the black experience per se and in the context of politico-economic dynamics, and given that environmental studies programs study, among other issues, the politico-economic factors affecting environmental quality and sustainability, it is crucial that the two work together. In short, I posit that the manipulation of racial suspicions and antagonisms is one of the most significant factors standing in the way of achieving a sustainable environment in the United States.

Environmental studies is to be applauded for its goal of generating knowledge for a sustainable economy and an overall sustainable human-environment interaction, a goal that would be impossible to meet with today's level of fossil fuel burning, which, according to recent reports, releases six billion tons of carbon dioxide into the atmosphere every year. The human, social, and environmental ills from this include undermined human health from air pollution, climate refugees, political instability resulting from global competition for finite oil deposits and other commodity sources, environmental degradation caused by global warming (e.g., more powerful storms, droughts, desertification, melting glaciers, and weaker ecosystems), acid rain, and water pollution.

A disturbing fact is that "through lawyers, lobbyists, elected officials, government regulators, conservative think tanks, industry front groups, and full-force media saturation, the [fossil fuel] . . . industry uses its wealth to change the public debate and, more often than not, achieve its desired policy outcomes" (Juhasz 2008: 11). More specifically, "three out of every four lobbyists [who] represent oil and gas companies were previously members of Congress who served on the committees that oversee and regulate the industry, or worked for various federal agencies responsible for regulating the energy industry. . . . [This is a major reason that,] from 2002 to 2008, federal energy subsidies to the fossil fuel industry totaled more than $72 billion. Renewable energy subsidies during that same period were less than $29 billion" (Rifkin 2011: 158). Nevertheless, it is not inevitable that a fossil fuel regime will dominate America. If the industry is unable to get 270 electoral votes, at least 60 reliable Senate votes, or a president who backs fossil fuels, it will be politically constrained as it has not been since President Theodore Roosevelt broke up Rockefeller's Standard Oil Company in 1911.

Chief among the many recent and current fossil fuel–backed think tanks and front groups are the Global Climate Coalition, the Heartland Institute, the American Enterprise Institute, the Manhattan Institute, Americans for Prosperity, the Alliance for Energy and Economic Growth, and the American Coalition for Clean Coal. These groups provide a platform for technical and political climate change deniers to confuse the public about the scientific consensus that climate change is real and anthropogenically induced. Deniers work to convince public officials and the general public that climate change is unproven.

Africana studies needs to be engaged in this epic struggle in the way its forerunner traditions of thought were engaged in struggles when science was used as a tool for dominating political reasoning, particularly when race with respect to slavery and Jim Crow was involved. For example, pseudo-science was employed to argue that blacks' cranial capacity was smaller, making them inherently intellectually inferior. The same brilliance, cleverness, and tenacity with which Africana studies scholars fought against this pseudo-scientific reasoning can also be employed by environmental studies scholars as they engage intellectually and politically to defeat climate change deniers and their supporters.

It should be natural for Africana and environmental studies to work together as the hip-hop generation gives way to the climate change generation. Bakari Kitwana, author of *Hip Hop Generation* (2002) and *Why White Kids Love Hip Hop* (2005), identifies as the hip-hop generation blacks born between 1965 and 1984. Hip-hop was once considered the expression of a counterculture and its resistance to prevailing social ills. As it went mainstream globally, it began to be repackaged as a cultural platform for marketing products to young adults. This distortion is recognized most graphically by hip-hop's overindulgence in "gangsta-ism" and "video vixen-ism." It would be naïve to think that, in its original artistic forms and political objectives, hip-hop would have been "allowed" to flourish as a medium for communicating and mobilizing around the ills of society as they are painfully manifested in young people's lives. To prevent it from meeting its political objectives, hip-hop came under excruciating and relentless attack, which sadly has led to more pronounced stereotypical images of the black experience.

Kitwana notes that early hip-hop art forms (e.g., DJ-ing, graffiting, rapping) gave voice to young people suffering from "reckless abandonment" by adults in their lives and in society. Their frustration with incarceration, illicit and pharmaceutical drugs, poor education, social inequality, and often grinding poverty was expressed with a distinct attitude and explicit body language.

As the birth period of the hip-hop generation was coming to an end in 1984, at least according to Kitwana's formulation, the "climate change generation" was starting to be born, according to Mark Hertsgaard, author of *Hot: Living through the Next Fifty Years on Earth* (2011). Hertsgaard tags its beginning year as 1988, which was the year in which NASA's James Hansen, one of the world's eminent climate change scientists, testified before a Senate committee that "global warming had begun." The following year, Bill McKibben wrote the first popular book on global warming, *The End of Nature*. He refers to our new planet as *Eaarth*, suggesting that it is similar but different and will become increasingly so throughout the life of the climate change generation. In his words, "We have changed the atmosphere and thus we are changing the weather. By changing the weather, we make every spot on earth man-made or artificial" (1989: 58). As a result, it is believed that we are losing over one hundred species every day. Nature's response is not man-made; rather, it is based on artificial stimuli according to a built-in logic formulated over billions of years. We have little idea of what that logic is, which is why the response to what we have done to the planet appears increasingly chaotic. Rather than responding to this transformative change as if it were linear, we should think of it and so respond to it as nonlinear climate chaos.

Some scientists now think that it was during the 1980s that humans began to overshoot the capacity of the planet. They believe that we are currently overshooting by as much as 150 percent its biocapacity to sustain the global economy and absorb our waste without losing its regenerative capacity. In other words, our industrial systems are overstressing our ecological systems, destroying biodiversity, exhausting resources, and dumping pollution on the planet and into its atmosphere faster than either Earth or Eaarth can absorb it. If the global economy were to grow by, say, 3.2 percent annually until 2050, we could overshoot Earth's carrying capacity by 500 to 700 percent. That means that several additional planets would be needed to maintain our lifestyle using today's technology.

The climate change generation will have to deal with much hotter cities, more frequent and violent mega-storms, rising seas, reduced agricultural yields, changing precipitation patterns and levels, and, maybe most intimidating, water shortages. They also must deal with co-optation and corporate manipulation by faux green washing and the urgent necessity to both mitigate the causes of climate change and adapt to the effects of environmental chaos. Undoubtedly they will butt heads with the fossil fuel industry.

That said, it is not inevitable that the fossil fuel industry will win out over the climate change generation. Some victories against the industry are already being won by the Sierra Club's Beyond Coal campaign. Since 2007, the groups affiliated with this campaign have succeeded in blocking some 130 new coal-fired plants through cancellations and prohibitions in some areas. As a result, Wall Street has downgraded many coal company stocks, and it is now difficult for the companies to secure financing and insurance. Beyond Coal campaigners are not anti-energy; they simply want carbon-neutral, renewable energy to replace fossil fuels. And they are getting their way. In 2008 and 2009, the world invested more in renewable energy than in fossil fuels or nuclear power.

The climate change generation is learning from previous generational movements, and they know that their movement is the largest the world has ever known. One of the beauties of it, is that it is leaderless, which means that no one person or group is available for the climate destroyers to "buy off," "drug up," or "wipe out" and in this way end or even derail the movement. Africana studies, which provided stocks and streams of knowledge for the hip-hop generation and studied hip-hop in all its genres and nuances, can contribute to the climate change generation's struggle against fossil fuels and toxic chemicals.

Although race, rather than class, is a far better predictor of the location of hazardous environmental sites and economic activities, the glaring exception of Appalachia comes to mind. According to Michele Morrone and Geoffrey Buckley (2011), poor Appalachian whites have mortality rates higher than those in a growing number of developing nations. For well over a century, they were treated like colonial subjects of an oppressed periphery country or like internal colonial workers in a wealthy core country—paid low wages and laboring in hostile working conditions to extract commodities that were shipped to more privileged regions, where they contributed to higher-value-added activity that created higher wages, better working conditions, and higher standards of living for those citizens.

As Morrone and Buckley report, community-based organizations (CBOs) and social movement organizations (SMOs) such as the Keepers of the Mountain Foundation and Mountain Justice are working to improve the environmental, working, and residential living conditions in Appalachia. Africana studies and the environmental justice movement should be working with these groups. As Oliver Cox stated in his classic study, *Caste, Class, and Race*, "[Blacks] and the poor whites are exploited by the white ruling class, and this has been done most effectively by the maintenance of antagonistic attitudes between the white and the colored masses. Could anything be more feared, is any aspect of race relations more opposed by this ruling class than a *rapprochement* between the white and the colored masses?" (1948: 473).

According to a recent study, we are not quite at a white-black rapprochement in Appalachia. Using the new Google Insights search tool, Seth Stephens-Davidowitz (2012), compared Americans' Google searches with their voting patterns to determine the frequency of words searched in different parts of the country. His study revealed which region had most frequently searched topics that included the word "nigger(s)." It turns out that West Virginia, which is ground zero for Appalachia, was number one. As progressive Appalachian CBOs and SMOs successfully battle racial ignorance and racial prejudices and as Africana studies successfully generates innovative knowledge and continues its efforts to undermine racism, there will be greater opportunities for more citizens from different walks of life and social environments to collaborate in reining in the power of the fossil fuel industry. Grace Lee Boggs eloquently but steelily asks, "How are we going to build a twenty-first-century America in which people of all races and ethnicities live together in harmony, and European Americans embrace their new role as one among many minorities constituting the new multiethnic majority?" (2011: 30). Africana and environmental studies are leading the scholarly, activist, and professional endeavors that are bound to push society forward in these areas.

Collaboration is the best way for Africana and environmental programs to meet the needs of multiple communities. For example, restricting coal-fired power plants lessens the need for coal and thereby reduces the environmental degradation from coal mining and preshipping production in Appalachia. And since coal-fired power plants, which emit arsenic, chromium, manganese, beryllium, and other toxic heavy metals, are typically located in black and other communities of color, the health of residents can be improved. Finally, since there are far more jobs generated by each megawatt of power from renewables than from fossil fuels, blacks, Appalachians, and other less advantaged groups will ultimately benefit from renewable energy and other forms of green job creation. Significantly, "in the wind industry alone, over 80,000 jobs have been created over the past decade—the same number of jobs that exist in the entire US coal mining industry. And wind still makes up only 1.9 percent of the US energy mix, while coal accounts for over 44.5 percent of US energy production" (Rifkin 2011: 43).

Coal-fired electric power as a share of the nation's total continues to decline; currently it stands at 36 percent. In March 2012, the Environmental Protection Agency (EPA) announced an upgrade to its restrictions to no more than one thousand pounds of carbon emissions per megawatt hour, which further weakens coal production (Barringer 2012). Coal producers had already been weakened by numerous environmental groups working with the Beyond Coal campaign. More green communities of color and a greener society can be achieved faster and in greater abundance when Africana and environmental studies work closely together.

Layout of the Book

Chapter 1 develops and examines the proposition that Africana studies—an important field with much to celebrate in nearly fifty years in the academy—has given insufficient attention to the environment. Its laudable mission, through scholarship and community engagement, is to be a transformative force in combating the many ills of the black community and in raising the prospects for improvements in the black experience. And it has contributed to the achievement of these goals except for one glaring omission: the environment. The environment in many black communities in the United States, in Africa, and in parts of the diaspora is literally killing black people, and yet only a tiny percentage of Africana professors and programs have the environment as a centralized focus of their teaching.

Chapter 2 takes on the challenging task of initiating a conversation between Africana and environmental studies that could lead to organic relationships whereby each has an important and prominent role to play in the other's endeavors. The two fields have long, proud, and unique traditions, but their missions in the academy and in broader society can be made more successful by collaborating intimately and endlessly. The conversation starts at the paradigmatic level in each field. I provide a brief primer on major paradigms, first in environmental studies and then in Africana studies. Then I illustrate how constructs associated with each paradigm in each field can connect with others. I provide only a sketch of the fields' multidirectional paradigmatic discourse, which will undoubtedly become much more sophisticated in due course. Also, I present a green Africana studies curriculum and its rationale.

In Chapter 3, I discuss locally unwanted land uses (LULUs), which are hugely disproportionately located in black and other communities of color, even when controlling for socioeconomic status. The LULUs closely examined are brownfields and toxics release inventory (TRI) facilities. In addition to LULUs, African Americans, who primarily reside in urban areas, are disproportionately adversely affected by urban heat island (UHI) effects. UHI "is defined as the temperature differential between the contiguous rural area and its related urbanized space" (Johnson and Wilson 2009: 420). Differences in the types of land coverage and the built community make it difficult for inner-city neighborhoods to dissipate heat at night in comparison to neighborhoods in outer suburbs and contiguous rural areas. People may live in the same metropolitan area, but they can experience different temperatures. Low-income communities and communities of color tend to live in hotter neighborhoods with detrimental health effects (Huang, Zhou, and Cadenasso 2011). Extreme heat and air pollution trigger significant health problems, which are expected to be further exacerbated by climate change (Harlan and Ruddell 2011). Africana scholars who do not engage environmental concerns might better appreciate how the physical environment in a great percentage of black neighborhoods is contributing to premature deaths, high morbidity rates, stunted human potential, and economic stagnation. Environmental scholars who have not focused on environmental problems that concentrate in neighborhoods of color may also benefit. I make a novel effort in this chapter to associate the hundreds of U.S. cities that have Africana studies programs with the brownfields and TRI facilities in them. Because the EPA arranges the United States into ten separate regions, I also aggregate the numbers of Africana studies programs and LULUs within each. There is a strong correlation between the locations of Africana programs and the major LULUs covered in this book. Among the important takeaways from this is that Africana studies programs need not leave their backyards to address environmental problems. As they more demonstrably engage local environmental problems, their students will gain additional interest in and expertise to take up such pursuits professionally.

Chapter 4 focuses on green jobs. African American and other students in Africana studies are similar to students in other academic programs in that they want their education and credentials to lead to interesting, fulfilling, and rewarding careers. In this chapter I illustrate the many opportunities for well-trained, credentialed professionals to address the industrially induced environmental problems of today—climate change, weakening ecosystems, and declining biodiversity among many others—or to contribute to the designing and building of the green economy and society of the future. Such a range of jobs will exist not only for Africana studies graduates but also for college graduates in other disciplines, of all colors, as well as those who are not college educated. I also illustrate the extent to which the green jobs of today are located in the states and EPA regions where Africana programs are located.

Chapter 5 explores environmental challenges and opportunities on the African continent. It also discusses the economic transformations that are well under way in many African countries. Africa is growing at twice the economic rate of Brazil and has a larger middle class than India. More important, childhood death rates have plummeted over the past decade. While all of this is obviously good news, there is a bit of a conundrum: many want quality of life to continue improving and for Africans to have more choices in their lives, but the gains that come from fossil fuel–based industrialization can be short-lived as they exacerbate Africa's environmental problems, which are manifested in droughts, floods, desertification, and weakening ecosystems and in the cascading problems that they trigger. I discuss how green African transnationalism may help African nations take on pioneering roles in niched green technology fields by making strategic use of the African diaspora, particularly in core nations.

Following each chapter are one or two vignettes written by individuals whose work and associations are already at the intersection of Africana and environmental studies. These practitioners and intellectuals are doing interesting and important work, and they add a texture to the discussion that would have been difficult to achieve without them. I think readers will find the vignettes stimulating and a source of ideas from which to draw for their own work.

Finally, I should point out that discussions of LULUs (especially brownfields) and green jobs are fraught with ambiguity. Academics, of course, can handle ambiguity, as it is typically associated with their starting points, and they work toward more certainty and clarity as best they can. Reaching for precise figures, or at least formulating clear and exacting principles while dealing with ambiguity, is the hallmark of much academic work, and patience is needed if one is working with imprecision. Thus, I cannot speak precisely about numbers of brownfields and green jobs, but I do want to work with the best numbers possible in my analysis of both. In other words, I do not avoid analyzing hugely important phenomena simply because of the ambiguity they embody. In Chapter 3 I discuss why the number of brownfields cannot be precise, and in Chapter 4 I do the same with green jobs. The pointedness of the conclusions I draw does not diminish because a given EPA region or state has a little more or a little fewer brownfields than the number I use in my analysis. Moreover, I use secondary data from an authoritative source: the EPA. This point is also relevant to green jobs. Consider, for example, that a commercial airline pilot and a pedicurist fall into pretty clear-cut job categories. This is not the case with green jobs. As I do for brownfields, I simply analyze job or employment data culled from an authoritative source—in this instance the U.S. Department of Labor.

The empirical data in Chapter 1 provide precise figures on the extent to which Africana scholars, programs, and academic journals cover the intersection of the environment and the black experience, but once again, there are limitations. I discuss these limitations and the methodology used to obtain the data. Note that the number of Africana studies programs given in Chapter 1 is different from the number cited in Chapter 3 because of differences in criteria and updates to *eBlack Studies*. Reproducing my study with the same methodology or with a totally different methodology could move the numbers a little higher or lower, but it would not alter the conclusion that Africana scholars have not covered the environmental concerns confronting the black community in the same way they have brilliantly covered other concerns—not even close. A goal of this book is to provide direction for further work that could be accomplished through the collaborative efforts of Africana and environmental studies scholars and professionals.

1/ Greening Africana Studies

Redemption, Redevelopment, and Remuneration in the Black Community

Scholarly contributors to and advocates of Africana studies are rightfully proud of the field: it has been successfully established in the academy for forty-seven years and has trained Ph.D.s in the field for twenty-seven. That Africana studies is flourishing is evidenced by its eleven doctoral programs and its high job placement rate of 75 to 100 percent (West 2012), which is higher than that in many STEM (science, technology, engineering, and mathematics) fields. Throughout its history, Africana studies has contributed groundbreaking knowledge about social dynamics and culture and, as a result, has influenced other disciplines and interdisciplinary programs.

However, rather than relish the long-fought struggles and successes of Africana studies, this book critiques an important omission—the environment—which has significant ramifications for the field and broader society and therefore must be addressed. Throughout the world, African peoples suffer disproportionately from environmental degradation, yet the academic field that was created to address their ills has given insufficient research attention to this fact. Relatively speaking, both globally and nationally blacks have gained the least from Western-led industrialization while in many ways suffering the most from its negative effects.

Since its founding, Africana studies has generated and mobilized university resources to solve real-world problems. The mantra at its founding was "Bring the campus to the community to serve the needs of the community." This chapter seeks to establish empirically that burning environmental issues and concerns have been overlooked in Africana studies and to demonstrate the imperative of correcting this omission expeditiously and effectively.

I acknowledge that environmental issues are likely taken up in courses with titles that may not suggest such coverage, and I hasten to add that Africana studies is hardly the only scholarly field that has given insufficient attention to the environment. This chapter reports that the track records of American studies and women's studies are not significantly better. Moreover, it is not just these fields; many others have also given sustainability short shrift. In fact, we could take a giant step back from the academy and focus on how the U.S. government and other principal institutions of society have overlooked the environment, and we are all poorer and more vulnerable for it in terms of climate chaos, environmental degradation, biodiversity loss, acidifying and rising oceans, weakened ecological systems, and other unsustainable phenomena.

Nevertheless, because of the exceptional importance of the environment, I submit that it should receive more direct attention, both governmental and scholarly. As the United Nations Environment Programme (UNEP) states, "Our environment is our life and when it changes for the worse, our lives change for the worse" (2005: 70). With that statement in mind, I analyze how Africana studies has not yet performed up to its brilliance in serving as a major generator of new knowledge to be used by grassroots movements, social movement organizations (SMOs), and public policy makers in dealing with the black community and its environmental concerns.

With few exceptions, thousands of Africana scholars and the field's three leading journals—*Journal of Black Studies*, *Black Scholar*, and *Western Journal of Black Studies*—have contributed immensely to an understanding of the conditions, interpretations, and agency of holistic and transforming black experiences. However, only an infinitesimally small percentage study the behavioral and biological impacts of locally unwanted land uses (LULUs) in the black community; how the community resists, remediates, and redevelops LULUs; or, now that we are possibly on the cusp of an industrial paradigm shift to green, how the black community can take advantage of new economic and career opportunities.

In the next section, I survey how the black community has suffered disproportionately from LULUs in recent decades, and I discuss the absolute necessity of forthright remediation as the economy begins to shift from a "destructo-industrial" paradigm to an "eco-industrial" paradigm. As I have stated elsewhere:

> State-of-the-art industrial technologies and production are so wasteful, toxic, and resource-depleting [that] we may as well refer to the hitherto industrialization as the "Destructo-industrial Age," which could not be more different from its likely successor, the "Eco-industrial Age." Destructo-industrial technologies are those technologies that focus on enhancing labor productivity with little if any regard for consequential harm to environmental sources and sinks. Conversely, eco-industrial technologies are those that focus on enhancing natural resource productivity by increasing performance and recycling with continuously less material usage. (2007: 146–147)

On this issue, James Gustave Speth, President Carter's environmental advisor and former head of the United Nations Development Programme (UNDP), states, "My conclusion, after much searching and considerable reluctance, is that most environmental deterioration is a result of systemic failures of the capitalism that we have today and that long-term solutions must seek transformative change in the key features of this contemporary capitalism" (2008: 9). Speth also argues that "the question for the future, on the economic side, is how do we harness economic forces for sustainability and sufficiency?" (11). The mission, the history, and the uniqueness of Africana studies are too important for the field not to be indispensable to the world's green future, whose framework for development is just now being constructed.

Peoples of African heritage throughout the world need to prepare themselves by acquiring the knowledge and skills to critique existing environmental degradation, contribute to plausible green futures, successfully compete for coming environmentally sustainable jobs, and qualify for future entrepreneurship opportunities. In the sections that follow, I posit that Africana studies can play an instrumental role in this preparation. Following that discussion, I present evidence supporting my thesis that Africana studies faculty, curricula, and journals have largely ignored the environment in the Africana world. Finally, I conclude with a discussion of missed opportunities and pathways for progress in the black community and broader society.

Environmental Remediation and Redevelopment in the Black Community

Africana studies was conceived as an intellectual force for informing activism that would drive social change and thereby help both bring about an end to racial oppression and initiate true racial equality. As one of the newest fields to be established in the academy since sociology a half-century earlier, Africana studies was arguably the most progressive academic field at the time of its birth. As the leading edge, it paved the way for other progressive fields, such as Latino studies, women's studies, Asian studies, Arab studies, third world studies, and environmental studies. The academy, the black community, and broader society have all benefited for nearly fifty years from Africana studies' involvement.

In many ways, environmental studies now makes far-reaching contributions to society similar to the way Africana studies made contributions decades earlier. Given the black community's disproportionate suffering from pollution because of its proximity to LULUs, and given that part of the founding mission of Africana studies was to draw attention to and to inform agency to eliminate deep-seated structural racial discrimination, one might think that the environment would have played a much more prominent role in Africana studies. For example, in the early 1990s few professors or journals documented the myriad health-related issues associated with three of the five largest commercial hazardous waste landfills in the United States. These sites accounted for approximately 40 percent of total commercial landfill capacity and were located in overwhelmingly African American or Hispanic communities (Lee 1992). History shows that humans have always buried and burned their refuse. However, in the industrial age trash became a problem because of both toxic and greenhouse gas–enhancing technology and intense consumer demand for "new and improved" goods. Discarded rubbish kept increasing, becoming ever more biologically and ecologically lethal.

The twenty-five largest industrialized countries now produce well over 90 percent of the world's hazardous waste. They receive the "benefits" of industrial-age materialism, while others are expected to bear the cost when it comes to the pollution. Hazardous waste dumps and landfills have mushroomed in black communities, but people living in these communities have not passively stood by. Instead, a few have arisen to fight the privileges brought to others by race and wealth. This is one manifestation of the "not in my backyard" (NIMBY) principle. White privilege, the economic ability to move to environmentally safer areas, protection from restrictive zoning laws and political lobbying, and high social capital all converge to enable white America to systematically cluster commercial hazardous waste sites in communities of color. Robert Bullard concurs, saying, "Environmental racism combines with public policies and industry practices to provide benefits for whites while shifting costs to people of color" (2005: 91).

Broadly, according to David Pellow, "racism has been an organizing principle of the modern world system since the rise of European states" (2007: 37). We find, he says, a "collocation of people of color and environmental hazards" (15). Until the 1980s, American industrialists were content to dispose of toxic and other hazardous materials in neighborhoods of color. Burying and burning toxic waste in the United States' "internal colony"—building on the models of Stokely Carmichael and Charles Hamilton (1967) and Robert Blauner (1972)—was simply cheaper and more efficient.

In general, race-related theories (Cox 1948; Winant 2000) have not paid much attention to environmental inequality, and environmental studies has not paid much attention to racial issues. For Robert Bullard, "In the United States, race interpenetrates class and creates special health and environmental vulnerabilities. People of color are exposed to greater environmental hazards in their neighborhoods and on the job than are their white counterparts" (1993: 26). However, between 1979 and 1989, the environmental justice movement began to crystallize and subsequently influence public policy and industry practices. For instance, in Northwood Manor, East Houston, African Americans in 1979 successfully stopped a garbage dump headed for their predominantly black, poor, and working-class community. A few years later, in 1982, African Americans in North Carolina's Warren County challenged public policy and commercial decisions to place a highly toxic polychlorinated biphenyls (PCBs) landfill in their community, which was 84 percent black and one of the state's poorest.

These "isolated" cases reaffirmed anecdotal insights that environmental racism was rampant across the country, but there was no authoritative study documenting such a systemic proposition that could be used to command public attention and influence policy. That began to change in 1983, after Congressman Walter Fountroy, chairman of the Congressional Black Caucus at the time, commissioned the General Accounting Office (now the Government Accountability Office) to study the topic and issue a report. The investigation culminated in *Siting Hazardous Waste Landfills and Their Correlation with Racial and Economic Status of Surrounding Communities* (General Accounting Office 1983). Although somewhat hesitant, the report largely confirmed the black community's suspicions of systematic environmental racism. Four years later, the United Church of Christ released a study that was even more pointed and damning, *Toxic Wastes and Race* (Chavis 1987), which clearly established a link between race, class, and hazardous waste sites.

The publication of *Toxic Wastes and Race* "did for people of color and the environmental justice movement what *Silent Spring* (1962), documenting sources of environmental pollution, did for the middle class whites in the 1960s" (Taylor 2005: 100). Black and other communities of color became more organized around race and pollution. In 1990, Greenpeace released a study documenting that the portion of the minority population in communities with incinerators was 89 percent higher than the national average and that property values in those communities were 38 percent lower than the national average. In 1991, people of color organized the First National People of Color Environmental Leadership Summit in Washington, D.C. The following year they established the People of Color Environmental Groups Directory as an organizing and networking tool for grassroots groups focusing on the joint concerns of environmental and social justice. By the mid-1990s, the environmental justice movement had begun to influence national public policy. In 1994, President Clinton signed an executive order directing federal agencies to implement strategies for identifying and addressing the unequal burden of negative environmental effects borne by communities of color and low-income populations. With this and other policy shifts supported by the environmental justice movement, it should have been expected that industry would fight back on multiple fronts—including lobbying to slow-walk enforcement of Clinton's executive order until it could be reversed and supporting scholarship refuting the link between race and the location of environmental hazards, or at least muddying the waters. Another tactic was environmental blackmail in communities of color—threatening loss of job opportunities if blacks and other people of color refused to accept *more* than their share of the environmental burden.

Despite a few studies to the contrary, scholarly evidence of the race-waste link is overwhelming, equal to the scientific consensus that anthropogenic climate change is real and has real consequences. Those who question this conclusion are, to me, about as intellectually and evidentiarily irrelevant as the flat-earthers, but they are consequential in that they have been successful in confusing the public on the issue and on the necessity of advancing beyond fossil fuels.

In major cities across the United States, the average poor or non-poor person of color lives in a more polluted part of town than the average white person does (Lester, Allen, and Hill, 2001). The Environmental Protection Agency (EPA) in a 2005 Associated Press report showed that African Americans are 79 percent more likely than whites to live in neighborhoods where industrial pollution is suspected of harming the health of the neighborhood's residents (Bullard 2005). In 2007, the United Church of Christ Commission for Racial Justice published a follow-up to its 1987 study that helped trigger the environmental justice movement (Bullard et al. 2007). Its principle finding was that race was the most significant factor determining the distribution of hazardous waste sites—trumping household income and home value. Of the forty-four states where hazardous waste facilities exist, forty have disproportionately high percentages of people of color living within three kilometers of the hazardous sites, and in these neighborhoods, 56 percent of residents are people of color. In neighborhoods where waste facilities are clustered, 69 percent of residents are people of color. According to the EPA, the five most prominent hazards in such neighborhoods are lead, waste sites (landfills, incinerators, and waste treatment facilities), air pollution, pesticides, and wastewater—primarily from inner-city sewers.

The collocation of people of color and hazardous sites is principally the result of the public policy and industry practice of targeting these communities; the residents of these communities do not choose to move into areas fraught with environmental challenges (Mohai and Saha 2006; Pastor, Sadd, and Hipp 2001). James Hamilton (1995), after studying commercial hazardous waste facilities' expansion plans between 1987 and 1994, came up with a power-base explanation: the companies behind these plans targeted minority communities because white communities would likely raise the company's location costs given their high social capital and efficacy in opposing unwanted development.

The environmental justice movement is helping communities of color insist on NIMBY in their own locales. According to Dara O'Rourke and Gregg Macey, "Community groups and non-governmental organizations across the United States have . . . begun to promote citizen participation as a means of improving environmental monitoring" (2003: 385). As an example, because environmental regulatory agencies in Washington, D.C., and in states around the country have been slow to respond to allegations of commercial violations of ambient air quality standards, since the mid-1990s "bucket brigades" have been forming to monitor and facilitate the enforcement of environmental laws. As O'Rourke and Macey explain, "Bucket brigades are groups of residents who live in industrial zones and are recruited to monitor air, using low-cost grab samplers, near oil refiners, chemical factories, and power plants" (2003: 385).

In the early 1990s, the environmental justice movement in the United States, marked by coalitions of civil rights, social justice, and environmental movements, began to gain in strength and scope. In response to its effectiveness and to the growth of similar organizations, such as the Global Anti-Incinerator Alliance, which works with local residents throughout the world to restrict where companies and governments can build new noxious facilities, domestic industrial polluters began to think and move beyond an "internal toxic colonial" model to a "third world toxic colonial" model. Burying and burning toxic waste in an internal colony in the United States was seen simply as cheaper and more efficient. But with communities of color suddenly and loudly invoking NIMBY, industrial polluters began to take the next path of least resistance: targeting Africa, Latin America, the Caribbean, and South Asia. According to Pellow, "Every nation in Africa has been hosting obsolete pesticide waste stockpiles. These are sites where pesticide drums are stored for at least two years, in eroding containers, allowing the toxins to leak into the surrounding ecosystems" (2007: 169). Offshore toxic waste trade is one of the negative consequences of a politically powerful domestic environmental justice movement and clearly illustrates the necessity of not only national but also transnational SMOs and thus a transnational environmental justice movement to head off a global race to the bottom.

Africa is only responsible for about 2 percent of the global carbon footprint, but Africans suffer the most from the effects of climate shocks. These effects are manifested in increased flooding in some parts of the continent and in droughts in others. They are also seen in increased rainfall, temperature, and humidity, which account for increasing cases of malaria cases. Malaria snuffs out the lives of roughly nine hundred thousand Africans every year, nearly 90 percent of them children. Climate shocks are also undermining agricultural production, which accounts for approximately 40 percent of Africa's GDP and nearly 60 percent of the population's livelihood. According to a UNDP report, "In Sub-Saharan Africa, the areas suitable for agriculture, the length of growing seasons and the yield potential of food staples are all projected to decline" (2007: 99). As the climate continues to worsen, the quality of life for the most vulnerable Africans worsens in lockstep.

Despite their critical importance, Africana studies has ignored public policy and industry practices that collectively amount to climate injustice and environmental racism in Africa, black communities in the United States, and other parts of the diaspora. Robert Bullard puts it succinctly: "Blacks must become more involved in environmental issues if they want to live healthier lives" (1990: 15). However, Africana programs have missed opportunities to (1) shape public policy debates on the environmental concerns of the black community; (2) produce needed studies on environmental affairs in local black communities; and (3) play a prominent role in the cultivation of environmental justice activists, including public intellectuals, scientists, lobbyists, and grassroots organizers. Green Africana programs are not merely the province of social scientists in the area of public policy or of scientists and engineers in the area of environmental discovery and technology; they are also producers of creative works that come to them from the humanities. Humanities scholars are perhaps best prepared to produce creative works that incite and agitate the public into action. Creative scholars in Africana studies can, for instance, through artistic realism, stir emotions about the nexus of the environment and the black community by exploring and transgressing boundaries between real and numerous alternative worlds.

Opportunities for Redevelopment and Remuneration

Understandably, the environmental justice movement initially focused on preventing the exposure of low-income communities and communities of color to the dangers of hazardous waste and on preventing that exposure from worsening for those who were already victims. While these efforts continue, the movement has broadened its agenda to include remediation and redevelopment of brownfields, which primarily exist in poor and inner-city communities of color. Brownfields may not pose the demonstrable and immediate health risks that active hazardous facilities pose, but their adverse effects are stealthier. Brownfield buildings may be structurally unsound and a threat to anyone in the immediate vicinity. Moreover, they often foul groundwater and are subject to the migration of contaminants. In general, brownfields undermine social capital, erode community pride, and stall economic development.

The EPA estimates that there are more than half a million brownfields in cities across the country. Their redevelopment is financed by Community Development Block Grant (CDBG) funding, Section 108 loan guarantees, Brownfields Economic Development Initiative (BEDI) grants, Economic Development Initiative (EDI) grants, and Renewal Communities/Empowerment Zones/Enterprise Communities grants. In partial response to the success of the environmental justice movement, the EPA in 1994 established the Brownfield Job Training Act to help prepare local residents for employment in remediating their communities. The Clinton administration went even further in 1997 by creating an interagency brownfields task force combining the resources of fifteen federal agencies to make a greater impact on brownfield initiatives. A National Partnership Action Agenda was created to promote cooperation among state, local, and federal governments; the private sector; and nongovernmental organizations (NGOs). Such programs sound impressive, but they are woefully inadequate, largely because their funding is totally incommensurate with the scale of the brownfield challenge.

There are economic and employment opportunities in cleaning up the five million acres of brownfields across the country. Meta-studies suggest that the average brownfield remediation and redevelopment project creates about ten jobs per acre (Howland 2007), which works out to fifty million jobs. Of course, that many jobs will not be created, particularly in light of average public investment per job—well over $14,000—which works out to $700 billion. After surveying nearly 230 EPA brownfield pilot-grant stakeholders, Deborah Lange and Sue McNeil (2004) concluded that long-term job creation and expansion of the local tax base are better predictors of success than the environmental cleanup itself. There are many important unanswered questions about brownfield remediation, such as whether local residents near the distressed sites will gain access to the new jobs and whether residents in the pre-redeveloped areas ultimately will be priced out of the community through gentrification. Additionally, there is the question of whether the negative externalities of the post-redeveloped brownfield community, such as crime and stigma, will still exist, say five or ten years down the road.

To put it baldly, does brownfield remediation and redevelopment actually work for some of the nation's most distressed neighborhoods? Admittedly, in the eyes of many public officials, answers to such direct questions do not inspire enthusiastic support for ramped-up investment in brownfield projects, but not investing in them is not an option. Remediation and redevelopment may not be an immediate precursor to job creation, but they are necessary if redevelopment of the brownfield community as a whole is ever to occur.

Community development organizations such as Sustainable South Bronx (SSBx) know that their communities do not have the luxury of entertaining these questions merely academically. They know that the biological, ecological, economic, and psychological health of a brownfield community is being undermined, and they are acting on the urgent imperative of remediation and redevelopment if residents are to have a higher quality of life and, simply, a longer life. Majora Carter, founder of SSBx, won a MacArthur Genius Fellowship by hitting two objectives with one effort: remediating one of the nation's worst contaminated communities and empowering the local population with skills to do it themselves. Africana scholars can bring their research skills to bear on the situation by studying gaps in knowledge, critiquing established lines of thought, suggesting alternative remediation and development strategies, and helping prepare Africana activist-intellectuals with the green sensibilities modeled by Ms. Carter and many others across the nation.

Most Africana programs are on campuses in urban areas, many near black communities with brownfield sites or other noxious facilities. However, local grassroots organizations and community development corporations (CDCs) may be working on remediation and redevelopment projects largely without the active involvement of a campus Africana program, its faculty, and its students. In addition to supporting local community-based organizations dealing with environmental issues and the national environmental justice movement, there is also a need for Africana programs to engage in public dialogue and help shape public policy not just on environmental issues in the black community, as important as they are, but more broadly on the future "green-collar" economy.

At this point in the destructo-industrial economy—which the Western world has had for two centuries—we are ready to start surrendering to an eco-industrial economy, which is based on biomimicry.[1] This means that we should seek inspiration from nature to design and redesign industrial systems that are sustainable and environmentally friendly—that is, decarbonized, detoxified, and dematerialized

[1] Biomimicry has been explained well elsewhere. For an introduction, see Hawken, Lovins, and Lovins 1999.

parallel forms of production. We should not be Pollyannas about the timing of the transformation, as it will not happen overnight. Nevertheless, all industrial and commercial shifts generate "winners" and "losers." The winners tend to be those who are agile and early adapters, while the losers are late adapters or even nonadapters. African Americans were slower at adapting to the digital economy than other groups, which further exacerbated established economic gaps. Abdul Alkalimat and Kate Williams (2001) attempted to engage Africana studies in educating the black community as well as public and corporate officials about courses of action to prevent the black community from falling further behind. Alkalimat and Williams demonstrated how Africana studies could establish community technology centers near campuses in poor or working-class black neighborhoods to provide residents with digital skills.

For a national workforce that has been buffeted by offshore production—from blue-collar assembly and textile work to white-collar engineering design and financial services—the notion of green-collar work that cannot be offshored is reassuring. Many green-collar jobs involve weatherizing homes and retrofitting buildings, which could be performed by workers formerly employed in manufacturing and digital industries.

Moreover, many of the white-collar and blue-collar occupations now outside the green sector will eventually be in it, even though entirely new job categories have been created, and many others will follow. As it stands now, many of the most skilled jobs cannot be filled, according to the National Association of Manufacturers (NAM). In a 2005 NAM survey, 90 percent of respondents indicated a moderate to severe shortage of qualified, skilled manufacturing and production employees such as machinists and engineering technicians (Eisen, Jasinowski, and Kleinert 2005). And when we look at the high-tech green sector, we see that the gap is expected to grow even larger.

Table 1.1 lists several white-collar occupations that play prominent roles in the green sector ("lime-collar" jobs), including the renewable energy industry. These include operations managers; electrical, chemical, and environmental engineers; and production managers. Average hourly 2007 wages in Ohio are provided for each. There are also "turquoise-collar" occupations, which are traditional blue-collar

TABLE 1.1 AVERAGE WAGES FOR LIME-COLLAR AND TURQUOISE-COLLAR OCCUPATIONS, OHIO, 2007

Lime-collar		Turquoise-collar	
Occupation	**Hourly wages ($)**	**Occupation**	**Hourly wages ($)**
Operations manager	41.30	Millwright	35.45
Construction manager	39.21	Electrician	22.20
Electrical engineer	38.45	Machinist	16.39
Chemical engineer	37.34	Roofer	16.31
Industrial production manager	36.94	Welder	15.26
Environmental engineer	35.45	Electrical equipment assembler	12.96
Chemist	28.62	Production helper	10.92
First-line production supervisor	23.23	Agricultural worker	9.43

Note: Figures adjusted from Pollin et al. 2008.

jobs in the green sector. Table 1.1 lists several of these—some skilled, others not—that have prominent green-sector roles to play. Turquoise-collar workers include electricians, machinists, roofers, production helpers, and agricultural workers. The table also provides the average 2007 wages in Ohio for each.

There is now a clarion call for Africana studies to shed light on how the emerging green economy can assist in raising the status of the lowest socioeconomic communities. Van Jones, former president of Green for All, notes, "The green economy should not be just about reclaiming thrown-away stuff. It should be about reclaiming thrown-away communities. It should not just be about recycling materials to give things a second life. We should also be gathering up people and giving them a second chance" (2008: 14).

It is safe to say that President Obama's environmental record has been mixed (Eilperin 2013). This is disappointing to active environmentalists who have expressed deep concern about the breaching or near breaching of planetary boundaries. Evidently, they see the president as being too cautious in his efforts to rein in the abuses of industrial polluters. They had expected that his environmental agenda would be guided more by the rectitude implicit in his phrase, borrowed from Martin Luther King Jr.'s 1963 "I Have a Dream" speech, "the fierce urgency of now" (see Klein 2011). Somehow these environmentalists, given their support of Obama, did not think that his presidency would by this time have facilitated a third increase in domestic oil production or that the United States would surpass Saudi Arabia as the world's largest liquid fuel producer. Also, at the time of this writing, President Obama has yet to rule on the Keystone Pipeline. In view of the environmental issues surrounding the pipeline, no doubt he will be judged harshly if he supports its completion and begins to transport the dirtiest form of crude with the lowest energy return on energy invested (EROEI), which is extracted from Canada's Boreal forest, the world's largest area of terrestrial carbon sequestration.

On the other side of the ledger, President Obama has introduced some significant regulatory impositions on fossil fuel industries and at the same time has enhanced renewable energy industries. For example, by 2025 the U.S. standard for auto production fleet miles per gallon will have doubled, to 54.5, from what it was when the president was first elected to office. Though this represents a significant reduction in gasoline use, it does not necessarily represent a significant reduction in comparison to what is needed. Furthermore, the increase in MPG over such an extended time frame does not put much pressure on the automotive industry, which did not strongly oppose the new standard.

At the end of 2011 the EPA imposed regulations, the first ever, on mercury emissions and other poisons from some 1,400 fossil-fuel power stations. In 2013 the administration tightened restrictions on carbon dioxide emissions, resulting in a reduction of over 35 percent and thereby limiting emissions to 1,100 pounds per megawatt hour for coal-fired facilities and 1,000 pounds per megawatt hour for large natural gas–fired facilities. Additionally, new coal facilities can be built only if they are able to capture and store some 30 percent of their emissions. Rather than just making the blanket declaration that no new coal-fired electric plants can be built in the United States, which would have engendered more cries of the president acting as a "socialist dictator," Obama set science-based emissions standards for production, which will likely have the same effect of limiting, if not preventing, further coal-fired electric plants. When we look at this regulatory policy in combination with the Sierra Club–led Beyond Coal campaign, we see that the president's impact on the dirtiest form of electric production is decisive and tangible. A final high mark on President Obama's environmental scorecard is that renewable energy production doubled during his first term in office because of the substantial increases in direct subsidies and tax credits he steered toward the industry. In part because of his policy agenda and price signals, investment in renewable energy exceeded investment in fossil fuel sources in 2009, and at this time, it is safe to say that more renewable energy support will come during his remaining time in office.

The administration's revitalization programs and, more broadly, its long-term economic agenda seek to mobilize public and financial investment in alternative energy and a greening economy. However, despite this great transformation, which is just getting under way and is affecting the entire planet, major Africana journals are not publishing research articles on the subject, Africana professors with an academic interest in the environment are not being appointed, and Africana studies courses that take up environmental issues in the black community are obviously not being taught.

Prior to the environmental justice movement of the early 1990s, public policy and industry practice took the path of least resistance and constructed hazardous facilities in black and other communities of color with impunity. Now that residents are becoming informed and vigilant regarding regulatory and planning matters concerning the environment, polluting commercial enterprises are finding it increasingly costly in those communities to build and/or expand noxious facilities. Communities of color have expanded their focus to include brownfield remediation and community redevelopment. And now that the gears are slowly beginning to turn for the transformation into a green economy, blacks and others of color have reason to anticipate remuneration through green-collar employment and green-sector investment. Again, Africana studies—judging by its faculty, self-professed interests, journal articles, and course inventories—has been on the sidelines of these exciting developments.

Methodology on Ascertaining Original Data

Gathering the data to test and support this chapter's chief proposition involved use of the Internet resource *eBlack Studies* (www.eblack

studies.org), which lists both undergraduate and graduate Africana programs at U.S. universities, complete with links to course and program websites. Again, Africana programs are not the only academic area paying insufficient attention to environmental concerns. The role of the environment in American and women's studies was determined from program lists obtained from the American Studies Association's online guide (http://www.theasa.net) and the Artemis Guide to Women's Studies (http://www.artemisguide.com), respectively. The purpose of using these sources was to ascertain (1) the number of professors who have research interests in the environment and (2) the number of courses offered in Africana studies, American studies, and women's studies that relate to environmental issues.

The total number of Africana studies undergraduate and graduate programs available for analysis of the number of faculty with interests in the environment was 240. Of the 219 total undergraduate programs in Africana studies, 159 were included. Community college programs and programs that offered only a minor, a concentration, or a certificate were excluded. Twenty-one programs did not have faculty interests listed on their website, and so these faculty were contacted individually, both by telephone and e-mail. Of the twenty-one programs, five were excluded because of nonresponse. Because Africana studies programs are frequently interdisciplinary, affiliated professors who had research and/or teaching interests in environmental issues may have come from outside the program; nevertheless, they were included in the data. Of the 2,253 faculty members we counted for inclusion in our analysis, a mere 2.8 percent, or only 63, identified environmental interests.

This 2.8 percentage does not square with Robert Bullard's resonating clarion call: "Blacks must become more involved in environmental issues if they want to live healthier lives" (1990: 15). Brilliant Africana professors with interests in history, literature, and public policy understandably stimulate students' imaginations and consequently inspire a number of them to take up careers in such broad areas. On the other hand, when these professors have no significant professional interest in the environment students miss out on the opportunity to become informed and excited about career options that relate to the environment and the black experience.

The number of environment-related courses offered to students was determined by a review of online course catalogs as well as course listings available on department websites. Out of thousands of courses offered to students in Africana studies, a mere twenty-eight undergraduate and six graduate courses on environmental issues were offered. (One in seven of those was in the program at the University of Toledo.) Similarly, as with faculty interests, this number included courses outside of Africana studies that students could take to fulfill related social science, natural science, and humanities degree requirements.

A comparison of the results from Africana studies to those in American and women's studies provided a random sample of all undergraduate and graduate degree programs—including every seventh program in the sample. Community college programs and programs that offered only a minor, a concentration, or a certificate were excluded. This resulted in a total sample of 23 American studies and 32 women's studies programs. Only 10 of the 248 American studies faculty members had research interests in the environment (4.03 percent) compared to 36 out of 714 women's studies faculty members (5.04 percent). With only 4 percent of American studies professors and 5 percent of women's studies professors having an expressed interest in environmental concerns, this result was similar to the 2.8 percent of environmentally focused faculty in Africana studies. With respect to course offerings, for American studies programs only 18 environment-related undergraduate courses were offered; only 9 were offered in women's studies. For the sample, there were zero such courses offered at the graduate level. Akin to the results from Africana studies, this number included courses offered outside of the department that students could take to fulfill outside social science requirements.

Following the discovery that Africana studies scholars are not identifying environmental issues as part of their repertoire of research, I expected to find that the three leading Africana journals—*Journal of Black Studies*, *Black Scholar*, and *Western Journal of Black Studies*—are also not dedicating adequate attention to environmental issues. Support for this assertion was obtained by analysis of journal archives for the years 1998–2013. During this fifteen-year period, the environmental justice movement began to emerge and exert influence

TABLE 1.2 AFRICANA STUDIES JOURNAL ARTICLES ON THE ENVIRONMENT, 1998–2013

Journal	Environmental articles/total articles
Journal of Black Studies	7/503 (1.39%)
Black Scholar	1/297 (0.34%)
Western Journal of Black Studies	4/242 (1.66%)
Total	12/1,042 (1.15%)

on public policy, and more national and international attention began to be paid to environmental concerns. Accordingly, it was presupposed that environmental issues would have begun to gain scholarly attention.

The laborious process of data gathering consisted of careful examination of all article abstracts and keywords in all issues of all three journals; the results can be found in Table 1.2, which illustrates that relatively few articles dealt with environmental concerns between 1998 and 2013. Collectively the journals published more than one thousand articles over that period regarding myriad aspects of the black experience; however, only about 1 percent of them dealt squarely with environmental matters.

It is recognized that leading African American scholars, such as Robert Bullard (1990, 2005), who are concerned with environmental justice, published in other journals outside of these three. However, because of the tradition of these journals and their present readership, and given that the attention paid to the environment in relation to blacks needs to increase, more articles on this topic should be appearing. Obviously, however, quality scholarship on the environment must be submitted to the journals before it can be published.

Dorceta Taylor (2007) of the University of Michigan surveyed college students across the country in environmentally oriented science and engineering programs. These programs included agricultural sciences, environmental engineering, environmental science, social sciences, natural sciences, biological sciences, forestry, geography, and geosciences. Of the 1,239 students who participated in the survey, 7.9 percent were black, 9.1 percent were Hispanic, and 1.6 percent were Native American. Regarding African Americans, Taylor found that the ideal job to this demographic was employment in government environmental agencies, and black students were far less likely than others to report being willing to work for mainstream environmental organizations. Overall, students reported a preference for working in government environmental agencies and in teaching. Alas, the survey respondents, including blacks, reported less interest in working in environmental justice organizations. Africana studies could help in a couple of ways. First, if such programs appointed more faculty with environmental research interests and professional activities and taught more environmental justice courses, as well as more entrepreneurship, public policy, history, and humanities courses with a focus on the environment, more African American students would likely take up an environmental focus as well. Second, because of the heavy humanities and social sciences orientation of the Africana curriculum, students who enroll in green Africana courses, including environmentally oriented science and engineering students, may think more favorably about working for environmental justice organizations. And with the great economic transformation now occurring, in which many of the so-called best and brightest can no longer work on Wall Street and in other high-paying corporate positions, a great number of students are reevaluating their career priorities.

Table 1.3 lists older Africana studies doctoral programs (before 2008), the number of faculty with interests in environmental research, and the number of respective courses on the environment. It shows that there is a palpable need for African Americans to take up leadership roles in program development, lobbying, and research related to environmental remediation, redevelopment, and remuneration. Without an integration of Africana and environmental studies, African Americans either are being trained in a research agenda removed from environmental matters or are being trained in an environmental studies program and so are less likely to take up research projects related more specifically to the black community or more generally to the environmental justice movement.

Table 1.3 also lists older graduate programs in Africana studies, their faculty with an environmental research focus, and related courses taught. Here, M.A. and Ph.D. program assessments are equally distressing. Africana studies has gone from being arguably

TABLE 1.3 AFRICANA STUDIES FACULTY WITH ENVIRONMENTAL STUDIES INTERESTS

University	Environmental research faculty to total faculty	Courses on environment
Africana studies Ph.D. programs		
Harvard University	2:33	0
Indiana University	0:49	1
Temple University	0.11	0
University of California–Berkeley	0:13	0
University of Massachusetts–Amherst	0:11	0
University of North Carolina–Chapel Hill	0:6	0
Yale University	0:30	0
Africana studies master's and graduate certificate programs		
Columbia University	0:8	0
Cornell University	0:16	0
Duke University	0:19	0
Florida International	0:12	0
Louisiana State University	0:7	0
Morgan State University	0:2	0
North Carolina AT&T	0:26	0
Ohio State University	1.47	0
State University of New York–Albany	0:10	0
University of California–Los Angeles	1:12	1
University of Iowa	0:16	0
University of Kentucky	0:3	0
University of Michigan	1:62	1
University of Nebraska–Omaha	0:9	0
University of Texas–Austin	0:7	0
University of Wisconsin–Madison	0:8	0

the most progressive field in the academy to one giving little attention to an area that not only has had a punishing impact on the black community but is on the cusp of a potentially great transformation: from destructo-industrialism to eco-industrialism. Even business schools have green MBA programs. Africana studies should be developing progressive green curricula.

In 2015, Africana studies will celebrate its forty-seventh anniversary in the academy and its twenty-seventh year as a doctoral field. Given the nature of racism in America and the efforts of established disciplines in the earliest years of its existence, many "enlightened observers" did not think that Africana studies would survive. However, not only has it survived, but its progressive push in the academy has opened the way for several other progressive programs, such as women's and third world studies, that together with Africana studies have changed the academy; more important, they have helped change America. Changes include everything from decentering "great white men" in history to academy-community engagement with and empowerment of working-class communities of color. Despite this venerable and storied history, in recent decades Africana studies has lost its progressive edge when it comes to the environment. Similarly, it has failed to take up new areas such as the Internet and other information and communications technologies. Some twenty years after becoming a part of public life, not to mention promoting career opportunities, there are still hardly any Africana professors who specialize in cyberspace or who offer such courses, and it is still somewhat rare for articles on these topics to appear in the field's journals.

This chapter illustrates that Africana studies has fallen short when it comes to environmental matters in the black community and in broader American society, if not in the global political economy. When the civil rights and social justice movements joined with the environmental movement to expand NIMBYism to mean not just affluent white communities but all communities, progress was made. But when polluters responded simply by engaging in toxic trade with cash-strapped nations of the Global South, transnational social movement organizations such as the Global Anti-Incinerator Alliance joined with local populations to expose and successfully challenge polluters in their burying and burning of toxic wastes among people who had few resources with which to resist. Progress in the promotion of life-supporting environments in black communities is being held up in part because of the disengagement of Africana studies on this subject in Africa and throughout the diaspora.

Expanding NIMBY beyond the white community to communities of color in the U.S. North and South and the entire Global South leads to NOPE—Not on Planet Earth. NOPE places more pressure on public and private funders of research and development to devise production processes and products that will lead us out of the destructo-industrial age and into the eco-industrial age, which is based on biomimicry and decarbonized, detoxified, and dematerialized forms of production.

Africana studies can be instrumental in fostering this transformation. Presently, it gives little attention to larger environmental questions in the same way that environmental fields often give little attention to developmental concerns in the black community. Green Africana programs can address both, and when they do, more African Americans are likely to enter careers and participate in social activism focused on green endeavors.

Green Africana Studies Curricula

Both Africana studies and environmental studies are known for offering latitude in course requirements while maintaining a core body and tradition of thought. Should the two become integrated to produce green Africana studies, we might expect even more latitude. The most common precept of sustainability is that each generation should interact with the environment to meet its present needs without limiting the ability of future generations to meet theirs. This idea holds even with different conceptualizations of sustainability. Moreover, the sustainability precept as discussed in this chapter comports with the sensibilities of the Africana experience of prima facie fairness and the circularity of life. Students who complete a green Africana studies program will feel comfortable and prepared to excel in scholarship and in professional activities relating to broad aspects of sustainability and the black experience.

A green Africana curriculum prepares students to critically evaluate sources and human impacts of environmental problems and solutions to them. These problems are most often hugely complex, and the way we understand them results from public policy, cultural, business, ethical, sociopolitical, and scientific points of view. It is impossible to overstress the point that excellent critical thinking skills are paramount to sorting out environmental problems and opportunities. The intersection of Africana and environmental studies provides the vocabulary, perspective, and confidence necessary to bridge multiple disciplines as a way to effectively engage environmental problems and solutions. Probably more so than in most disciplines, Africana and environmental studies are comfortable with problem- and solution-centered curricula, since their raison d'être is to make a material impact respectively on the black community and on the environment. Green curricula focus on applied knowledge for solving real-world problems at the local, regional, national, international, and global levels.

Environmental studies is concerned with the relationship of humans to their ecosystems. But, as we know, on one level there are environmental problems and opportunities for all of humanity (e.g., global climate chaos), and on another level there are environmental problems and opportunities that are racially, class, and geographically specific. The greening of Africana studies is critical because it will keep the field true to its mission of improving the lived conditions of people of African heritage and becoming thoroughly knowledgeable of them. A specific point to note is that degraded environments are disproportionately killing and debilitating more African Americans than white Americans, and the effects of Western-triggered climate change are killing and debilitating more Africans than Europeans.

Again, if Africana studies is to be true to its mission, it must green its curriculum to prepare intellectuals and leaders with the interest, skills, and credentials to work through institutions—some of which they will help found—to improve the environment and, in turn, the lives of black people. Similarly, studying the black experience through Africana studies will help environmental studies fulfill its mission of educating humanity so that its necessary interactions with the environment will safeguard our fragile climate system; protect ecosystems; preserve nonrenewable resources; and prevent toxins from undermining the biological health, economic viability, and social capital of communities, particularly those of people of color, who are disproportionately affected by environmental ills. Environmental programs with an Africana point of view can do the work of preventing and positively reversing the environmental fortunes of such communities, which are among the most vulnerable.

In Table 1.4, the first two columns, respectively, list courses that might very well constitute the fundamentals of Africana and environmental studies, recognizing that these two decidedly interdisciplinary fields afford broad latitude within traditionally recognized core curricula. In other words, no program in the country will likely have all of the courses identified, but they will no doubt have fairly close facsimiles. Each program starts with a gateway course and concludes, as

TABLE 1.4 AFRICANA, ENVIRONMENTAL, AND GREEN AFRICANA STUDIES CURRICULA

Africana studies	Environmental studies	Green Africana studies
• Intro to Africana Studies • Intro to the African Experience • Cultures of Sub-Saharan Africa • African Politics and Governance • Research Methods in the Black Community • African American History • Black Intellectual History • African American Writers • African Literature • African American Cultural History • Senior seminar • Capstone project	• Intro to Sustainability • Environmental Social Science - Environmental Policy - Environmental Anthropology - Sociology of Sustainability - Environmental Politics - Environmental Economics • Environmental Humanities - History - Philosophy and Ethics - Literature - Aesthetics • Environmental Science - Earth Science - Geology - Green Chemistry - Biology - Ecology - Physics • Human Ecology • Cultural Ecology • Energy and Society • Statistics • Senior seminar • Capstone project	• Intro to Africana Studies • Intro to Sustainability • Research Methods in the Black Community • Intro to the African Experience • Environmental Social Science • Environmental Humanities • Environmental Science • Environmental History and Challenges in U.S. Black Communities • Environmental Communications • Environmental History and Challenges in Africa • Environmental Inequalities and Opportunities • Ecotourism Studies of the Africana World • Environmental Policy • Sustainability Entrepreneurship • Sustainable Business Practices • Geographic Information Systems for Sustainability • Statistics and Environmental Data Analysis • Natural Resource Management • Industrial Ecology • Sustainability Internship • Senior seminar • Capstone project

indicated in the table, with a senior seminar and a capstone project. Africana students learn, at a minimum, African American and African history, culture, and politics; black intellectual history; and Africana perspectives and research methods. In environmental studies, students take, at a minimum, requisite environmental social science and humanities, environmental science, and human ecology courses. The social science, humanities, and science fields require students to enroll in perhaps two or three courses in each. Although the social sciences and the humanities are emphasized, environmental science and quantitative reasoning are hugely important, particularly for purposes of professional competence and the self-confidence necessary to challenge polluters and their apologists, and as a way to attain credibility. There are likely ideal or aspirational courses in both Africana and environmental study programs, but what is actually offered is often dictated by the talents and interests of faculty.

A green Africana studies curriculum will integrate some of the fundamental courses found in current or traditional environmental studies. Given that sustainability will be only one theme of a green Africana studies program, a shared course distribution obviously will not be 50-50 with environmental studies. Instead, Africana courses will predominate and likely incorporate carefully selected environmental courses, as identified in the third column of Table 1.4. Africana programs with a bigger commitment to sustainability undoubtedly will have more green courses.

Essentially, a green Africana studies curriculum will sacrifice depth for breadth in its green courses. This should not be surprising in light of its broader mission, which is to improve economic, political, social, and other lived conditions in black communities based on knowledge of their sources. Thus, green Africana studies curricula should include courses such as Natural Resource Management, Geographic Information Systems for Sustainability (environmental informatics), Industrial Ecology, Sustainability (Green) Entrepreneurship, and Sustainable Business Practices in addition to some fundamental liberal arts environmental courses. Enrollment might require partnering with, for instance, business and natural resources programs. Green Africana should also include an internship and a capstone project in which students will take the comprehensive knowledge and technical skills they have acquired and relate it to a black community or to broader communities. The project would be evaluated primarily by academics.

An internship would provide an opportunity to work in a public, nonprofit, or private-sector institution that addresses environmental problems or pursues environmental opportunities. Consistent with other types of internships, students would apply their disciplinary and interdisciplinary knowledge and skills to real-world problems. Students would be evaluated at the end of the internship by both academics and institutional practitioners.

The discussion in this chapter represents merely the scaffolding at the intersection of Africana and environmental studies. The next chapter is a richer conversation on this topic. Curriculum scaffolding is institutional, whereas theoretical perspectives within fields of study are both institutional and personal. They are institutional in that they have historically, if not transhistorically, evolved into traditions of thought exemplified by individuals who become a given field's most notable members—for example, W.E.B. Du Bois and Rachel Carson, noted for their activism and their shaping of, respectively, Africana and environmentalism as fields of study and activism. Contemporaries of each historical movement use and contribute to established and institutionalized traditions.

Theoretical perspectives are personal in that they speak to the individual; separately or in combination they resonate with people, helping them make sense of the world through their field of study. Exposing Africana students to environmental studies in a unified context will enable them to appreciate such a synthesized perspective and empower them to address environmental problems and pursue environmentally oriented opportunities in black communities nationally and internationally.

The remaining four chapters focus on integrating perspectives from Africana and environmental studies (Chapter 2), environmental problems in the black community (Chapter 3), green job opportunities for the black community (Chapter 4), and environmental problems and opportunities in Africa (Chapter 5). The Conclusion weaves together the various strands of the book. In addition, I have selected several individuals of diverse backgrounds to write short vignettes about their green intellectual and practical work in black communities; these vignettes appear at the end of each chapter and help bring to life some of the chapter's major points.

VIGNETTE 1.1. EDWINA KOFI-OPATA

Africana Studies Student Develops New Passion for the Environment

Although I was aware of global warming, climate change, and issues pertaining to the environment as a whole when I first started college, my knowledge of the environment was minuscule. I was, however, given a wider platform when I decided to write on some aspects of climate change within the international system as part of my thesis requirement in graduate school at the University of Ghana. I became intrigued with how developing countries were positioning themselves to tackle head on the issue of global warming and climate change and the problems they created for societies. My curiosity had been aroused following the research, but I did not know what I was going to do next. Everything was put into perspective when I came to the University of Toledo for another graduate degree. Although my major at the time was political science, I knew I wanted to focus on the environment—but from a global perspective, not one limited to a particular geographic region. I began to look for courses that were geared toward my interests.

I registered for an Africana studies class entitled Environmental Inequalities and Opportunities, though originally I was not sure what it held in store for me. All I knew was that the class was going to address the environmental inequalities faced by sections of global society based on race, class, and location. My experience in taking the class was overwhelming. My professor constantly emphasized the opportunities to be gained by moving away from the use of fossil fuels to clean energy and encouraged us to investigate further. Based on that encouragement, I applied for an internship with the Columbus, Ohio, chapter of the Sierra Club. Although I wanted to be at the forefront of discussions on renewable energy, I was assigned to researching coal plants. The Sierra Club has been in the vanguard of promoting an effective moratorium on new coal power facilities through its Beyond Coal campaign. I was fascinated by the social dimension of the research: the underlying questions of how communities in which coal plants existed were involved in the process of shutdowns and the unnerving contention surrounding the future of coal plant employees, as well as alternative uses for degraded lands. The debate in recent times has been on how to cope with the massive job cuts that would result should global society transition from the use of fossil fuels to renewable energy. However, the green economy would provide just as many jobs as the status quo. To bridge the gap of unemployment, emphasis should be placed on skills development at different levels of engineering targeted at coal plant workers in particular as well as the general public. Engineering skills are needed to support various aspects of the green economy such as the building and servicing of wind turbines, solar panels, and other forms of technology. The ready availability of individuals with the necessary skills is an important and necessary building block. One interesting observation in the course of my research was the alternative use of remediated brownfields that generate income for the site communities and address the issue of lost taxes.

On the whole, what I gathered from my research experience was an awareness of the constant need for communities to be proactive in every activity within their domain—particularly with regard to the environment. More broadly, my internship work has connected with my earlier interest in developing countries and their position within the international system as it pertains to global warming, climate change, and the transition to a green economy. As that transition continues, further research into satisfying the energy needs of the African continent will be essential. With the completion of my master's degree behind me, I am now a doctoral student in spatially integrated social sciences. My focus is on energy and environmental policy with an emphasis on energy sustainability in Africa.

2/ We Have a Lot in Common

Let's Talk

Environmental Studies and Africana Studies Paradigms

In all academic fields of study, multiple explanatory and predictive paradigms compete for dominance. Scholarly works tend to fit in a given paradigm or a school of thought, and rarely do scholars switch one for another in a serial manner. The major Africana studies paradigms covered in this chapter are class analysis, Afrocentrism, Africana womanism/black feminism, and radical egalitarianism. The major paradigms of environmental studies are ecological modernization, environmental justice, eco-Marxism, ecofeminism, and ecocentrism.

Paradigms, or schools of thought, are organized around a set of assumptions and propositions about various aspects of the subject area in question. Take African womanism as an example. Africana scholars who think in this paradigm ask questions and make observations regarding black women in relation to family, other black women, the black community, and broader society. Afrocentrism, another paradigm, asks questions and makes observations regarding the culture and psychology of black people living in a Eurocentric society. Divergent paradigms lead to different questions, and different presuppositions, concepts, and prescribed propositions explain how a field is generally structured and operates in the real world. However, they are nonetheless unified in their relationship to their field, which Abdul Alkalimat (2001) refers to as a paradigm of unity. All of the competing Africana studies paradigms may be used to explain any aspect of the black experience, but all are united in their purpose: to increase the knowledge needed to advance the global black experience.

It is hard to overstate the fierceness with which disciplinary paradigms compete. Some have even tried to "excommunicate" their competition, seeking to make their paradigm the only one. Progressive programs such as Africana, whose raison d'être is ending racial oppression and elevating the global black experience, and environmental studies, whose goal is ending environmental destruction and promoting sustainable societies, can have paradigms that are as inhospitable to one another as they are to the colossal social phenomena they seek to address.

This discussion of paradigms is not meant to break new ground. Instead, the five environmental studies paradigms and the four Africana studies paradigms are presented merely as a primer for those outside one or both fields. Thus, well-versed Africana scholars might skip the coverage of their paradigms, and environmental scholars might skip the coverage of theirs. Since the value of this book is its bridging of the two fields, the goal here is to enlighten scholars in other fields or those who are new to Africana and environmental studies about each field's paradigms.

Environmental Studies Paradigms

Environmental studies has multiple paradigms competing for dominance. The major ones are ecological modernization, environmental justice, eco-Marxism, ecofeminism, and ecocentrism. Though all of these schools of thought seek to advance knowledge of and to alter public behavior against environmental destruction, they can be as hostile to one another as they are to the humans and institutions that have demonstrably destructive environmental sensibilities and practices.

The most fundamental distinction between environmental studies paradigms is whether they are anthropocentric. Orthodox, or strong, anthropocentrism holds that humans have full dominion over all nonhuman species and the rest of nature (Norton 1984; Dobson 2007). Its roots lie in the reigning interpretations of Judeo-Christian beliefs (at least up through the twentieth century), modernity, scientism, and patriarchy. The values and norms of the institutions who adopted these interpretations with respect to the environment were hegemonic in that they created social-environmental interactions that were then broadly adopted as a matter of course. The polar opposite of anthropocentrism is ecocentrism, the perspective not only that humans are no more important than any other species but that nature has intrinsic value irrespective of human needs and beliefs. "Orthodox anthropocentrism" was associated with the exploitative capitalist approach to nature throughout American history up to Rachel Carson's publication of *Silent Spring* in 1962. As explained later, exploitative capitalism has been unapologetically, wantonly, and implacably destructive to nature and has been justified in terms of "development" in the interest of humans, particularly Europeans and North Americans.

Ecological modernization, environmental justice, eco-Marxism, and ecofeminism are anthropocentric, but they are more "neo" (i.e., weak) than the now fading orthodox paradigms. That is, they still privilege the interest of humans, but they are respectful of nature. William Grey (1993) discusses what he calls "enlightened anthropocentrism," which sees human-centeredness as natural, inevitable, and certainly potentially benign if we consider the interest of nonhuman species and avoid upsetting nature. Thus, although these four paradigms hold axiomatically that humans are the superior species, they recognize that we pay a price for abusing the environment, and so they also recognize the imperative of altering our behavior with respect to it.

Only ecocentrism counsels us to alter our environmentally destructive behavior because we are abusing the rights of other species, which are morally equivalent to humans. The first four paradigms, referred to as "neo-anthropocentrist," counsel us to alter our environmentally destructive behavior out of enlightened self-interest. All five paradigms advocate reform, although in different ways.

Ecological Modernization

Before ecological modernization was established as an environmental paradigm, there were two general attitudes toward the environment, which can be labeled the exploitative capitalist paradigm (ECP) and the romantic/transcendentalist paradigm (RTP) (Taylor 2000). Driven by wealth accumulation, the ECP countenanced the wanton destruction of nature throughout much of America's early history, which is not surprising given that exploitative capitalists also condoned slavery as a means of wealth accumulation. Biblical scripture interpreted nature as God's gift to humans for their use. Nature was also viewed as a female to be tamed and controlled by men for their pleasure. Conquering nature was regarded as the key to wealth and happiness, similar to the benefits of conquering other people.

The environment, according to the ECP, was boundless in terms of its richness and ability to absorb waste and destruction. The nineteenth-century clearance of America's forest represents the most destructive human-made environmental transformation the world had ever known ("The Best Story" 2012). Again, a similar view was held regarding the conquering of other people. Serious men of industry were about the business of controlling other men, nature, and women, whereas perceived weak and sensitive men concerned themselves with the well-being of others and nature, and regarded themselves as only equal to women. Of note, the ecofeminist perspective, discussed later, connects the domination of females and nature (Shiva 1989; Warren 1997).

In the late eighteenth century and throughout the nineteenth century, a period in which horrific destruction of nature was occurring because of frenetic industrialization, the romantic/transcendentalist paradigm (RTP) started to gain traction. Conservationists reacting to the Industrial Revolution included Ralph Waldo Emerson and Henry David Thoreau, who sought to conserve nature for humankind's aesthetic, spiritual, recreational, and inspirational appreciation, as evidenced in their respective publications *Nature* and *Walden*. John Muir, preservationist and cofounder of the Sierra Club, was even more ecocentrist than Emerson and Thoreau in that, from his vantage point, the wilderness was to be preserved independently from the value that humans may desire from it or ascribe to it.

Such romantic transcendentalists are believed to have given environmentalism its initial impulses of nonsocial justice, wilderness needs over human needs, and anti-urban bias (Rhodes 2003), which continued until the late twentieth century with the rise of the environmental justice paradigm. Those who approached the environment from a romantic transcendentalist perspective wrote about the intrinsic value of nature, its beauty and truth, its representational extensions of a supreme being, and its ability to expand our humanness beyond consumption of natural resources. For Thoreau, the pathway to understanding life on Earth was through nature, and with nature being raped, we were rapidly and violently losing our ability to understand ourselves. Thoreau's *Walden*, Emerson's *Nature*, George Perkins Marsh's *Nature and Man*, and Muir's *A Thousand Mile Walk to the Gulf* captured the reactions of conservationists and preservationists to the unapologetic and implacable onrush of the Industrial Revolution that was forever transforming both nature and humanity.

During this time sociology emerged as a discipline to better understand the social impacts and human toll of the Industrial Revolution and its attendant urbanization. Romantic transcendentalists and sociologists were horrified by the effects of the Industrial Revolution and intellectually curious about the ways in which it was permanently altering the trajectory of humanity. One key difference was that sociologists often believed that the excesses of industrialists could be tempered with systematic and rigorous scientific knowledge of society. The belief was that industrialists would be among those to see that it was in their enlightened self-interest not so much to stop industrialization and urbanization as to manage them more humanely and efficiently. Looking back, this view seems incredibly naïve. Nevertheless, it provided the intellectual basis for the ecological modernization paradigm. Still, the approaches of the romantic transcendentalists were regarded as woefully inadequate in dealing with the implacable industrial capitalism of the mid-nineteenth century. Their discourse was incapable of stopping exploitative capitalism and its frenetic environmental destruction, which is why they retreated. Ecological modernization stepped in to fill the void.

Ecological modernization seeks to resolve environmental crises through economic efficiencies and technological fixes. Thus, it imagines science and technology as the principal supporters of the environment rather than the principal culprits of environmental and social destruction (Mol 1996). Other paradigms, presented later, hold instead that technical solutions alone will not get us out of this vicious cycle (Dobson 2007). It is fair to say that ecological modernization has enjoyed hegemonic status in environmentalism in part because predominant institutions such as the government, giant corporations, mainstream media, other leading opinion makers, and formerly the general public all subscribe to ecological modernization in principle if not in name.

The ecological modernization paradigm is also considered to have the largest following considering that the "human exceptionalism paradigm" (HEP) and the "new environmental paradigm" (NEP) are a part of it. Simply, HEP comports with the religious and capitalist-materialist view of most Americans rather than with the ecocentrist view. Few Americans are going to believe that they are no more important than fruit flies, because humans alone can contemplate their mortality as well as ponder notions of a supreme being. Moreover, Americans may see themselves as more "Promethean" than others in light of their privileging of technological fixes for any problem, including environmental crises.

The NEP emerged in 1960s America as part of the post–*Silent Spring* environmental sensibility, and it has been prominent ever since. Rachel Carson's *Silent Spring* and the work of millions of citizens deeply moved by it led to the further demise of rather timid romantic reformist ideas of conservation and preservation. Actions to conserve nature for human recreation and aesthetic and spiritual appreciation, as well as for its intrinsic value independent of human needs, had been proven incommensurate with the challenges facing the environment. NEP and its adherents recognized that much more comprehensive and politically engaged action was necessary to extract the requisite concessions from industrial polluters and other orthodox anthropocentrists in the interest of human health and safety.

Ecological modernization is rooted in the social science school of modernization, which itself emerged out of evolutionary and functionalist theory (So 1990). It contends that societies progress unidirectionally: from traditional and simplistic to modern and complex. Accordingly, change is to be harmonious, centralized, top down, and aided by modern science and technology. The rationale for ecological modernization in the 1960s was that modernization thinking and planning had played a principal role in extending human life and improving its quality with safe drinking water, better shelter, and antibiotics; in shrinking the world via the Internet; and in exploiting the supposed self-correcting ability of science to reject false claims to knowledge. Therefore, according to this paradigm's perspective, modernization thinking and planning should be applied to the environment to address problems that are at least on the same scale as earlier monumental accomplishments.

An emergent area of ecological modernization is "industrial ecology," whose principal aim is the establishment of an industrial metabolism that is consistent with nature's metabolism. The idea is that industrial production should mimic ecology since the output of one species is simply nourishment for another, thereby leaving nature without waste. If industrial production is ever to achieve zero waste and pollution, its proponents believe, industrial ecology will be the means to do so. Paul Hawken, Amory Lovins, and L. Hunter Lovins (1999), major supporters of ecological modernization, contend that it can achieve a material and energy reduction of up to 95 percent in highly industrialized nations without sacrificing the quantity or quality of products and services because it pursues what is known as the "triple bottom line"—people, planet, and profits. If corporations are to be more successful over the long run, they must focus not just on profits but on social equity and the environment as well.

The World Commission on Environment and Development (known as the Brundtland Commission) has been a major proponent of industrial ecology. Its definition of sustainability is an environment that meets the present generation's needs without compromising the ability of future generations to meet theirs. Essentially all mainstream environmental nongovernmental organizations (NGOs) have adopted the ecological modernization paradigm. These include, among many others, the Natural Resources Defense Council, the Environmental Defense Fund, the Sierra Club, the National Audubon Society, and the National Wildlife Federation.

Environmental Justice

The ecological modernization paradigm may have the most adherents, but the environmental justice paradigm is the fastest growing.

Environmental justice is about fairness as social consciousness of environmental destruction grows. Climate change, depleting aquifers, wild weather patterns, and toxic environs are problems for all humanity, but some populations suffer first and worst from these and other effects of environmental degradation. All of the environmental studies paradigms address environmental ills that affect humans and nature, but only environmental justice focuses exclusively on the uneven distribution of those ills in communities of color and the poor. It contends that no racial, class, or any other social group should bear a disproportionate share of the environmental problems attributable in large part to industrialization. Further down on the mitigation and adaptation side of the ledger, no racial, class, or any other social group should be underrepresented in the transformation to a green economy.

Environmental justice has been the portal through which African Americans, other people of color, Appalachians, and other poor and working-class whites have come to reinterpret and support environmentalism where they had little interest in it before. Its launching was the period between 1979 and 1982. In 1979, six of Houston's eight garbage incinerators and all five of its landfills were sited in black neighborhoods, despite the fact that blacks made up only 28 percent of the city's population. The 82 percent black Northwood Manor near Houston filed a lawsuit to stop yet another landfill—Whispering Pines—from being sited in its neighborhood. In 1982, thousands of North Carolinians protested the dumping of polychlorinated biphenyl (PCB) in a landfill located in a section of town that was 75 percent black and in Warren County, which was the ninety-seventh poorest out of a hundred counties in the state. As a result of the protesting, over five hundred citizens were arrested.

A Government Accounting Office (GAO) study in 1983 demonstrated the correlation between hazardous waste siting and the race and class composition of surrounding communities. In 1987 an even larger study, commissioned by the United Church of Christ (UCC) Commission for Racial Justice, known as "Toxic Waste and Race," pointedly concluded that environmental injustice was not limited to places like Warren County; instead, it was a national pattern (Chavis 1987). African Americans had long held that their communities were singled out for environmental abuse, but they had only ethnographic anecdotes, not systematic, scientific studies as proof until the GAO and UCC reports appeared (Bullard 2007).

The popular view of African Americans and environmentalism is that blacks did not show any concern for environmental issues until the environmental justice movement came into being. Such a misreading of history is akin to concluding that blacks were not concerned with civil rights until the civil rights movement in the 1950s and 1960s. It is equivalent to suggesting that Africans of Zaire did not try repeatedly to remove Mobutu Sese Seko from office during the neocolonial period of the 1970s and 1980s, a suggestion ill-informed about the part played by American, French, Belgian, and other militaries and intelligence agencies in putting down organized uprisings. Zairians had been trying to dethrone Mobutu, but they were unsuccessful until 1997, when the West no longer needed him after the collapse of the Soviet Union.

In fact, blacks had been attempting to draw attention to degraded environments and the siting of hazardous facilities in their communities long before the emergence of the environmental justice movement, but the structure of society before 1978 did not afford them the institutional means to effectively fight against environmental racism.

Generally speaking, African Americans have never received justice when it comes to living free of disproportionate environmental degradation, an injustice that was primarily a benefit to other communities. Fortunately, the streams of social justice and environmental protection merged somewhat during the last two decades. The movement that blacks are joining and energizing today is environmental justice, which, as noted previously, holds that no community should bear a disproportionate share of the burden of an industrial society and that all communities should benefit from the transition to a global green economy.

Mainstream environmental organizations at one time acted as if all environmental positives (e.g., the wilderness for aesthetic, recreational, spiritual, and inspirational purposes) and all environmental negatives (e.g., hazardous waste, air pollution, groundwater contamination) were, respectively, equally accessible and distributed. Such a conclusion, drawn largely from a position of white privilege, explains why they refrained from analyzing environmental positives and negatives along class and racial lines. The unequal accessibility of environmental positives is not as immediately consequential as the unequal distribution of environmental negatives. In other words, blacks' inability to visit Mount Rainier National Park as an environmental good was less disruptive to their health than their inability to escape noxious air, soil, and water in their neighborhoods.

Blacks carried a disproportionate share of environmental negatives resulting from the sharp post–World War II uptick in industrial output. "Between 1946 and 1966 total utilization of fertilizer increased about 700 percent, electric power nearly 400 percent, and pesticides more than 500 percent. In that period the U.S. population increased by only 43 percent" (Rhodes 2003: 50). The "need" for another massive landfill near Houston was a reflection of the huge industrial upsurge in the 1970s, when the city was rapidly becoming the nation's fourth largest. It set Robert and Linda McKeever Bullard in motion to trigger the environmental justice movement. The unequal sharing of the pollution stemming from that upsurge was borne by African Americans, but since they were fundamentally without access to even the ballot box between 1946 and 1966 in the unapologetically racist society of the time, there was no political space for an effective environmental justice movement. However, a decade of civil rights victories resulted in sufficient structural reforms to allow its birth and expansion.

The environmental justice movement has galvanized millions of African Americans in response to the hazards and risks concentrated in their communities. The awareness and mobilization as a result of being better informed has significantly altered the equation, which means that polluters no longer have a free hand to degrade the environment in communities of color and low-income neighborhoods with impunity. While there is considerably more work to be done, the environmental degradation in predominantly black areas, once steadily increasing, is now actually shrinking. Again, this is not to suggest that degraded black neighborhoods are seeing remediation and redevelopment. Rather, the movement has had a material impact on the lives of millions of African Americans, other communities of color, and poor and working-class white communities all across the country.

It is also worth noting that the environmental justice movement has had a huge impact on environmentalism and mainstream environmental organizations and on government agencies with arguably the most direct interaction with the environment: the EPA and the departments of Agriculture, Energy, and the Interior. A common critique of environmental organizations, especially but not only from the corporate community and the political Right, used to be that they cared about things, not people. That is, they had an ecocentric focus rather than an anthropocentric focus. Environmentalists unwittingly seeded this thought by actions that seemed to privilege the wilderness and the rest of the Earth over humans. A common characterization of these organizations is that they have a neo-anthropocentric focus in that they strive to conserve nature for humans to behold for aesthetic, spiritual, and inspirational purposes. Meanwhile, the environmental issues and concerns of African Americans were essentially absent from the agendas of any environmental organizations before the establishment of the environmental justice movement as a progressive paradigm and movement for change in the 1990s.

Before the 1990s, environmental organizations had few African Americans on staff and next to none in their professional and decision-making ranks. Moreover, they initially ignored criticisms from environmental justice academics and practitioners about such omissions, if not their conscious discrimination against blacks. The absence of diverse staff was a moral failure as well as a failure to fulfill the organizations' missions. As the criticism mounted, environmental organizations initially became indignant about implicitly being called racists. This was a major turn of events because as environmentalists they fashioned themselves as among the most progressive and open-minded members of society; after all, they were speaking for those nonhuman species who could not speak for themselves. Next, they made excuses for having only a few blacks on staff and almost no black decision makers, such as that blacks were uninterested in environmental concerns (Sandler and Pezzullo 2007). Finally they came up with a *mea culpa* and simply got on with diversifying, which they have been doing for the past several years. Diversification requires more work in the professional and decision-making ranks, but there has been indisputable progress attributable in no small part to the environmental justice movement.

However, the contention of environmental organizations now is that it is exceedingly difficult to find blacks and other underrepresented groups with the education and credentials to add value and effectiveness to their missions. Because the major organizations appoint individuals with training in environmental sciences, national resource management, and other environmental fields, in which there are relatively few blacks, there is a relatively small pool from which they can draw. For this reason, Africana studies can play a major role in motivating African Americans through the education and credentials to work with environmental organizations and to start their own where needed.

Diversification has, in effect, led to a change in the DNA of environmental organizations. Nearly all of them now support the principles of environmental justice and, in varying degrees, have altered their infrastructure to advance its work. For instance, the Sierra Club now has the National Environmental Justice Grassroots Organizing Program, which was formed in 2000 (Sandler and Pezzullo 2007). Its Beyond Coal campaign, while forcing power producers to close down aging coal-fired facilities and switch to natural gas and renewables, is particularly beneficial for blacks and other people of color because coal-fired power stations tend to be located in their neighborhoods.

Environmental justice has also influenced federal agencies. The EPA was established in 1970 to protect environmental quality, but until the 1990s it gave little attention to the increasing degradation of environmental quality as communities became less white and less affluent. The environmental intellectuals, professionals, and grassroots activists who were inspired by works such as Rachel Carson's *Silent Spring* were not thinking about industrial degradation in relation to social class and racial status. Even the rightly celebrated *Silent Spring* gave no suggestion of what I call the second level of environmental inequality.

At the first level of environmental inequality, which Carson covered exclusively and brilliantly, a powerful set of industries (chemical, fossil fuel, manufacturing) causes significant environmental degradation from which it benefits and from which others suffer. There is certainly no justice in that equation. The energy behind limiting and mitigating injustice is what led to the EPA and to its mission and priorities. At the second level of environmental injustice, social groups with the least power and economic means are forced to suffer from the worst environmental degradation caused by industries with the most power and financial means and that receive the greatest benefits. In other words, environmental externalities are not evenly disseminated. The worst effects are borne by those with the least political influence.

In the early 1990s, the environmental justice movement began influencing the EPA, causing it to change toward the issues of race, class, and the environment, although progress has not been linear. The movement's impact on the EPA was more discernible during the Clinton administration, less so during the Bush years, and again more so during the Obama administration. There is no longer any doubt in EPA officials' minds that, if left to market forces and political expediency stemming from social inequality, communities of color and lower-income communities will carry a disproportionate share of environmental risks and hazards. The agency's embrace of environmental justice is further demonstrated by the fact that it was included in the EPA's fiscal year 2011–2015 strategic plan (see U.S. Environmental Protection Agency 2010).

In addition to, or perhaps as a result of, its acceptance of and responses to established environmental justice principles, the EPA now has a diverse workforce. Scholars and activists had consistently pounded the agency on this issue similarly to how they had demanded data from traditional environmental organizations to show that blacks and other people of color, particularly in professional and decision-making positions, were underrepresented. In 2006, blacks made up 17.5 percent of the EPA's total workforce of 18,240 and 15.3 percent of its senior professional ranks. (Chapter 4 presents detailed data on African Americans at the EPA and the departments of Agriculture, Energy, and the Interior.)

Edwardo Rhodes's observation, that "the EPA still is dominated, regardless of the party in power in the White House, by professionals who come from a culture that is wanting in sensitivity to social justice" (2003: 90), does not quite ring true for the Obama administration. Previous EPA heads and executives operated exclusively according to the paradigm of ecological modernization, as much of their task, while supporting industrial production, was to regulate and, hence, to mitigate industrial harm to the environment. However, under Lisa Jackson's leadership in the first Obama administration and that of Gina McCarthy in the second, there has been a marked change from previous administrations stretching all the way back to the agency's founding. Jackson was the first administrator to adopt the ecological modernization and environmental justice paradigms. However, to the dismay of many environmentalists, she is a modernist in support of industrial production without what they would call adequate industry regulation. At the same time, judging by her public remarks and policies, she is rooted in the paradigm of environmental justice, linking the environment and race, class, gender, and social justice in an explicit framework.

Eco-Marxism

Eco-Marxism is like ecological modernization and environmental justice in the sense that all three paradigms are anthropocentric in placing the needs of humans above those of other species but recognizing that it is in our enlightened self-interest to mitigate harm to the environment. The ecological modernization paradigm advocates mitigating, if not preventing, environmental degradation for all, nationally and globally; the environmental justice paradigm advocates mitigating, if not preventing, disproportionate environmental degradation when, if left to market forces and political expediency, it concentrates in communities of color and poor and working-class communities. The eco-Marxist paradigm advocates ending the type of capitalism we have known, if not ending it altogether, to achieve social equity and a sustainable environment. In other words, because eco-Marxism identifies capitalism as the root cause of environmental degradation and social inequality, it holds that ending wanton environmental degradation and achieving sustainability cannot happen as long as capitalism as we know it exists. Tim Jackson captures this dilemma: "The capitalist model has no easy route to a steady state position. Its natural dynamics push it towards one of two states: expansion or collapse" (2009: 64). And, as Albert Einstein famously pointed out, we cannot solve our problems with the same thinking we used when we created them. In other words, eco-Marxism is fashioned as the only anthropocentric paradigm that is outside the worldview of capitalism, so, as the paradigm postulates, it is the only one capable of advancing a sustainable economy.

The mind-set of capitalism is endless economic growth to sustain profits and employment. For this reason, according to eco-Marxism, capitalism must either grow or die. Economist Kenneth Boulding famously put a finer point on the problem: "The only people who believe in infinite growth in a finite world are madmen and [capitalist] economists" (quoted in Gilding 2011: 64). The paradigm of eco-Marxism conveys a sense of urgency. According to Joel Kovel's (2007) stark framing of this issue, either we stop capitalism from destroying nature or we allow capitalism to destroy the world.

Eco-Marxists believe that the environment and capitalism cannot sustain each other. The environmental crisis has to be addressed with new eco-technologies that bring an abundance of environmental positives to everyone (e.g., distributed solar and wind energy), though doing so threatens the productive relations of capitalism (i.e., the ability for a relative few to profit from satisfying mass needs using social labor). Either that or the advancement of such technologies must be impeded to protect those relations, even though this ironically undermines both capitalism and nature. One way or the other, as James Speth (2008), cofounder of the Natural Resources Defense Council, contends, capitalism will be fundamentally transformed, or it will be replaced by a qualitatively different economic system.

The specific issue receiving the most attention in eco-Marxism is the second principal contradiction of capitalism. The primary contradiction is social production and private appropriation. In other words, despite the fact that every commodity, product, or service embodies social labor or the labor of many, profits from the product or service's marketing are privatized. This contradicts labor's desire for more control of production and more benefits from it because corporations want to maximize the control of both profits and production. In the second contradiction, corporations seek to appropriate nature for private gain while the costs of environmental abuse are shared or externalized in common (O'Connor 1998). When a small number of industrial enterprises do this, they have a temporary advantage over their rivals, but when all of them devour nature for "continuous growth" using environmentally destructive technologies, then capitalism, humanity, and other species are put in peril. The cumulative effects of pollution, mineral depletion, climate change, and disruption of ecosystems are generating a new public consciousness of and clamor for renewable energy and other eco-technologies that, when broadly embedded in the economy, forever alter its character.

One reason for Marx's intense focus on productive forces—in addition to his view of them as crucial to explaining and driving social revolutions transhistorically—is that productive forces have to advance up to a given threshold to create a sufficient abundance of goods for the principle of "to each according to his/her needs" to work. However, on the basis of today's state-of-the-art technologies, the more production we have, the more environmental degradation we have. Some green advocates see this as a laudable social equity principle that assures all humanity of an acceptable minimum of life-sustaining essentials such as clean water and clean energy. However, the result is a blank check for technological advancement and its attendant environmental degradation because there are, theoretically, no limits to "needs" or to the "wants" that can be manipulated into perceived needs. Hence, a major problem with this entire formulation for many greens is that eco-Marxists are too willing to sacrifice the environment in the interest of false needs satisfaction. In other words, advancing endlessly to higher technological thresholds and meeting unlimited human needs are prescriptions for environmental collapse.

Eco-Marxists attempt to address this issue in two ways: by decoupling economic production from environmental degradation and by decoupling human happiness from human material consumption. John Bellamy Foster, a leading eco-Marxist, describes how the first decoupling works:

> Logically, in order to be physically sustainable, an ecohistorical formation has to meet three conditions: (1) the rate of utilization of renewable resources has to be kept down to a rate of their regeneration; (2) the rate of utilization of nonrenewable resources cannot exceed the rate at which alternative sustainable resources are developed; and (3) pollution and habitat

> destruction cannot exceed the "assimilative capacity of the environment." (1999: 132)

Meeting these conditions means advancing technologies to the point that they embody decarbonization, detoxification, and dematerialization of production or declining material throughput.

To achieve the sustainable future that Foster and other eco-Marxists imagine, technologies can no longer contribute to carbon emissions that lead to climate change, they can no longer be toxic to ecosystems or be nonbiodegradable, and they must rely primarily on recycling and declining levels of energy and material throughput. Achieving these economic and ecological outcomes requires both relative and absolute decoupling, as Tim Jackson describes:

> It's vital here to distinguish between "relative" and "absolute" decoupling. Relative decoupling refers to a decline in the ecological intensity per unit of economic output. In this situation, resource impacts decline relative to the GDP. But they don't necessarily decline in absolute terms. Impacts may still increase, but at a slower pace than growth in the GDP. The situation in which resource impacts decline in absolute terms is called "absolute decoupling." Needless to say, this latter situation is essential if economic activity is to remain within ecological limits. (2009: 67)

The second decoupling involves delimiting consumption. Eco-Marxists such as David Schweickart (2011) have pointed out that greater consumption beyond a surprisingly low threshold has scarcely anything to do with greater human happiness. Therefore, because consumption beyond a given threshold does not generate more happiness, it makes no sense to have more production of finite resources if the pollution it creates is absorbed by an environment with a finite capacity. Such thinking is, of course, inimical to capitalism. Recall the capitalist growth imperative of "grow or die." Truncating production beyond some aggregate happiness threshold—assuming it could be empirically determined and enforced—would be tantamount to truncating and ending capitalism. At the end of the day, as the

Nigerian poet and novelist Ben Okri observes, "we must bring back into society a deeper sense of the purpose of living. The unhappiness in so many lives ought to tell us that success alone is not enough. Material success has brought us to a strange spiritual and moral bankruptcy" (2010: 312).

Ecofeminism

The diversity within the paradigm of feminism includes, among others, liberal, Marxist, and radical perspectives. These perspectives give rise to debates about the sources of and solutions to sexism, but they agree on the mission: working to end it. Feminism is based on defeating "power-over" logic. In other words, the logic of domination has to be defeated for sexism to end. That same logic of domination originated and has sustained not only sexism but also racism, classism, and naturism (i.e., the domination of nature). Vandana Shiva (2010, 1989) and others contend that the destruction of women, nature, and subjugated cultures are all rooted in the logic of domination. The ecofeminist paradigm posits that the domination of women and the domination of nature are linked.

Ecofeminism is anti-androcentric but not necessarily anti-anthropocentric. Naturism was borne of males taking the lead role in violently transforming nature (e.g., felling forests, damming rivers, and commercializing agriculture) and generating metaphors that made them the sadistic male protagonist versus "mother" nature as the object of male domination (Collard 1988). Chauvinistically, "nature is often described in female and sexual terms: nature is raped, mastered, conquered, controlled, mined. Her secrets are penetrated and her womb is put into the services of the 'man of science.' Virgin timber is felled, cut down. Fertile soil is tilled and land that lies fallow is 'barren,' useless" (Warren 1993: xv).

Although ecofeminism is clearly a major opponent of the androcentric domination of nature that we have known throughout history, the ecofeminist paradigm is not necessarily an opponent of anthropocentrism. Karen Warren, a leading ecofeminist intellectual, argues that humans are superior to other species but also argues that that superiority is not to lead to the abuse of nature (Marina 2009). She uses the metaphor of adults and children: the former is "superior" to the latter, but the superior power is not to lead to adults abusing children. Moreover, Warren does not say—perhaps because it goes without saying—that on a moral basis adults should not dominate children, but when it comes to human superiority over nature, she suggests that abuse should not occur for reasons of enlightened self-interest. Warren would undoubtedly argue that adults and children have morally equal status, but she would likely not make that claim regarding humans and the environment.

The principles of ecocentrism, which is described later, are not those from which ecofeminism draws. Biocentric egalitarianism, a major tenant of ecocentrism, places humans on par with all other species. Warren and Val Plumwood (1997), another leading voice in ecofeminism, reject biocentric egalitarianism and embrace human superiority and centeredness, making ecofeminism by definition anthropocentric. Ecofeminism and ecocentrism are mutually exclusive: the former privileges the female as the center, while the latter privileges nature. Furthermore, ending sexism, one of the goals of ecofeminism, can be accomplished without embracing ecocentrism.

Table 2.1 shows the percentage of female employees in four federal entities that interact directly with nature: the EPA and the departments of Energy, Agriculture, and the Interior. It also shows the percentage of females serving as senior professionals and agency executives. We see that women have considerable ground to make up, most notably in the departments of Interior and Energy. A greater female presence in these departments would provide more career opportunities for women; also, a woman's wider perspectives on society-environment interaction would better serve all of society.

TABLE 2.1 WOMEN IN FEDERAL AGENCIES AND DEPARTMENTS MOST CLOSELY ATTACHED TO THE ENVIRONMENT

Agency	Total employees (%)	Senior professionals (GS 12–15) (%)	Executives (%)
EPA	50.6	47.5	34.0
Agriculture	43.0	36.2	25.7
Energy	37.9	36.6	19.1
Interior	40.6	34.5	27.3

The Neo-anthropocentric Paradigm

The neo-anthropocentric paradigms covered in this book—ecological modernization, environmental justice, eco-Marxism, and ecofeminism—offer distinct discourses and frameworks for understanding the society-environment nexus. At the same time there is noticeable overlap among them. All four want economic growth to continue. The differences are in who the primary beneficiaries and losers are and who controls the growth and output.

Ecological modernization wants to mitigate degradation so as not to threaten the very ecological foundation on which the global economic system rests. Elites of oligarchic capitalism have a stake both in the mitigation of degradation and resource depletion and in the protection of a system of growing economic inequality.

Ecological modernization theory, or EMT, "posits that environmental problems may be mitigated by increasing resource efficiency, improving sustainability, while retaining the basic system of capitalist production and consumption" (Zhu, Sarkis, and Lai 2012: 170). In responding to these issues, ecological modernization elites have established institutions such as the World Business Council for Sustainable Development and the U.S. Climate Action Plan.

Environmental justice proponents want economic growth to continue, but they do not want the most politically and economically vulnerable members of society to be burdened disproportionately with the health-undermining and environmental risks that come with it. Environmental justice differs from ecological modernization in seeking to politically empower and gainfully employ communities of color and low-income citizens specifically in greening the old industrial economy from which they were largely shut out and whose effects they have suffered.

Eco-Marxism also wants economic growth to continue in order for all humanity to benefit, not just industrial investors or targeted communities of color and low-income citizens. Since the rapacious appropriation of nature in today's global economy is based on the logic of capitalist accumulation rather than social need, to achieve ecologically rational production (O'Connor 1998), reigning global oligarchic capitalists will have to be replaced by social equity–minded and sustainability-minded stewards of the economy and the environment.

Deep ecology, an intellectual strand of ecocentrism, places ecofeminism in the same category of shallow ecology as it does the other environmental paradigms discussed. What this means is that ecofeminism is like ecological modernization, environmental justice, and eco-Marxism in that it is a neo-anthropocentric paradigm vis-à-vis ecocentrism. All four neo-anthropocentric paradigms unapologetically acknowledge that they are human-centered, not biocentric, and therefore they advocate the continuous mitigation of environmental abuse out of both enlightened self-interest and appreciation for nature's intrinsic value. Each privileges humanity and especially some parts of it (the working class, people of color and low socioeconomic status, and women) as beneficiaries of declining environmental abuse, which will come about by attacking not only resource inefficiency and pollution but also the uneven distribution of power wielded by white-dominated economic institutions and capitalist oligarchies and by supporters of patriarchy, both men and women. In other words, environmental justice, eco-Marxism, and ecofeminism are long-term projects with missions that require the defeat of racism, class domination, and patriarchy—the "power-over" logic of organized domination of others.

Ecocentrism

Among the paradigms of environmental studies, ecocentrism is the least applicable to Africana studies. Since the latter is about advancing knowledge of the global black experience, it has a thoroughly neo-anthropocentric orientation. Biocentrism and ecocentrism are the polar opposites of anthropocentrism. Biocentrism equalizes all life, thereby removing humans from their hierarchical perch over other species. Furthermore, the notion of equalizing moral concerns and obligations to support all living matter is simply alien to Africana studies—and most Americans—and is therefore at odds with philosophers such as John Dewey who do not accept ontological anthropocentrism, or the position that humans are somehow privileged in a cosmic sense. Urban African Americans are also likely to look askance at John Muir, who wrote, "Now my eyes were opened to [flowers'] inner beauty, . . . revealing glorious traces of the thoughts of God, and leading on and on into the infinite cosmos" (1913). Biocentrism is regarded as too ethereal to address the central concerns of blacks stemming from the legacy of enslavement, colonization, and Jim Crow, as well as from a racially hierarchical world: concentrated black poverty, dysfunctional predominantly black schools, struggling and transforming families, and transnational capitalists in a global economy that yields fewer pathways for social mobility.

In Paul Wapner's words, "If biocentrism puts life at the center of moral concern, eocentrism expands the moral boundaries even further to include all aspects of nature" (2010: 67). J. Baird Callicott (1987), an ecocentrist, contends that we have an ethical impulse and a sense of obligation to support anyone recognized as a member of our community. This contention is likely a by-product of the evolutionary psychology used to support self-preservation. African Americans' predominance in Africana studies is evidence of such an ethical and intellectual impulse regarding one's community.

Aldo Leopold is among the most celebrated ecocentrists. His writings aimed to convince humans to develop a kinship with morally equal fellow creatures, including the likes of fruit flies. The idea is to convince humans to extend their ethical sensibilities—beyond their community, beyond their racial-ethnic group, beyond their nation, beyond humanity, and beyond living things—to include all of nature (Pojman and Pojman 2012).

Whereas intellectuals such as Leopold and Callicott make the philosophical case for ecocentrism, others such as Dave Foreman and Arne Naess promote action to build an ecocentric movement. Naess (1973) coined the term "deep ecologists" to distinguish a group that claims concern for the environment for its own sake. "Shallow ecologists" are concerned with pollution, resource depletion, and energy efficiency, as well as the overall objective of good stewardship of the environment. Environmental scientists such as G. Tyler Miller (1972) contend that the Earth does not belong to us; instead, we belong to the Earth.

According to Naess's formulation, ecological modernization, environmental justice, ecofeminism, and eco-Marxism, the neo-anthropocentric paradigms, are all examples of "shallow ecology." Those who operate within them hold the view that since the environment has instrumental value to humanity, they are taking measures in varying degrees and through various approaches to protect it. Conversely, deep ecologists state axiomatically that all organisms have intrinsic value and an equal right to live and blossom. Deep ecology is largely the paradigm of privileged white males. It places no value on work done by so-called shallow ecologists—who are actually working to reduce ecological problems. In fact, for deep ecologists, shallow ecologists are no different from exploitative capitalists or orthodox anthropocentrists. Deep ecology presupposes that sustainability means radical changes in social and political life. Environmental problems are far beyond technological fixes and stem from political, economic and cultural relations that mutually encourage unsustainable practices. It is not a simple matter of reducing carbon content, toxicity, material and energy content per product and service, and overall throughput but rather a matter of less production—period.

Dave Foreman is a cofounder of Earth First!, an organization that, as its name states, places the environment over humanity, which has been destroying the environment since the industrial age began (see Foreman 1991). The activist wing of deep ecology sees shallow ecologists as, at best, slowing the momentum of humanity's destruction of the planet, but to truly reverse the course of the destruction requires colorful and urgent action to seize the attention of the general public so as to put pressure on industrial and public policy decision makers. For example, sympathizers of Earth First! have been accused of torching newly constructed "McMansions" and hummer dealerships to draw attention to the need to end excess. Clearly, Africana studies and ecocentrism are incompatible. Indeed, achieving the mission of one militates against the mission of the other. Africana studies does not have this problem with any other environmental school of thought.

Table 2.2 lists the environmental studies paradigms along with units of analysis, constructs, and example exponents in each case.

Africana Studies Paradigms

Many of the Pan-Africanist intellectuals and leaders who have influenced Africana scholarship and activism were influenced by Marxism. Although they did not embrace all of its tenets, they also did not dismiss it as Eurocentric or as having no relevance to black peoples'

TABLE 2.2 ENVIRONMENTAL STUDIES PARADIGMS

Paradigms	Units of analysis	Constructs	Exponents
Ecological modernization	Resource efficiency and waste elimination	Industrial ecology; triple bottom line; technological fix	Paul Hawken; Amoy Lovins; L. Hunter Lovins; Stuart L. Heart
Environmental justice	Environmental inequality; environmental remediation	LULUs; environmental racism; environmental and social equity	Robert Bullard; Dorceta Taylor; David N. Pellow
Eco-Marxism	Class domination to appropriate use of nature; privatization of natural resources; socialization of environmental degradation	Grow or die; treadmill of production; capitalism as equal to environmental destruction	John Bellamy Foster; Joel Kovel; James O'Connor
Ecocentrism	Biocentric egalitarianism	Authoritative/neo-anthropocentrism; deep/shallow ecology; intrinsic worth	Arne Naess; David Foreman; Aldo Leopold
Ecofeminism	Twin domination of females and nature	Power-over; logic of domination; naturalism	Vandana Shiva; Karen Warren; Val Plumwood

oppression. Rather, they culled the merits of Marxism that could be interpreted by and applied to black peoples' historical and geographical struggles. These intellectuals and leaders, who are in the pantheon of Pan-Africanists, are W.E.B. Du Bois, Kwame Nkrumah, C.L.R. James, Eric Williams, Amilcar Cabral, Frantz Fanon, and George Padmore. Other notables are Malcolm X, Assata Shakur, Cyril Briggs, Ella Baker, Walter Rodney, Claudia Jones, Victoria "Vicki" Ama Garvin, and Richard Wright. All helped build the intellectual foundation for black radicalism, which is defined here as philosophies and practices that articulate deep-level social transformations in the lives of black people, which require the dismantling of systems of oppression. Radical politics as defined by Winston James is "the challenging of the status quo either on the basis of social class [and] race (or ethnicity), or a combination of the two" (2000: 292).

The twin systems of oppression for Pan-Africanists are white supremacy and imperialism. The absence of women in the Pan-Africanist intellectual pantheon may partially explain why patriarchy was not originally considered a system of oppression as well. However, black radicalism is not static, and patriarchy has now joined white supremacy and imperialism to make a trilogy of oppressive systems to be dismantled. And Africana studies, which embodies black radicalism, can meet this mission only by producing knowledge to address the trilogy through its four paradigms: class analysis (i.e., black Marxism), Afrocentrism, Africana womanism/black feminism, and radical egalitarianism.

As Rose Brewer (2003), Carole Boyce Davies (2008), and Denise Lynn (2002) point out, the black radical tradition as formulated by Cedric Robinson (2000) is incomplete with only Pan-Africanism and class analysis. For them, what is missing is Africana womanism/black feminism.

Class Analysis

Pan-Africanists who seek a better understanding of the black experience to prepare themselves to dismantle the systems of African peoples' oppression often recognize the indispensable utility of classical Marxism.[1] The Marxist paradigm provides not only a vocabulary

[1] The class analysis paradigm is largely consistent with black Marxism. Although Marxist and black Marxist thinking exists today and is highly applicable to today's socioeconomic developments in American society and throughout the world, it may or may not be labeled as Marxism or black Marxism. (Major tenets of Marxist analysis often appear without the latter label.) In fact, some of the growing inequality of today cannot be explained adequately without incorporating some analytical strands of Marxism, at the core of which is class competition, exploitation, and domination. Black Marxism has focused on how class exploitation is expressed as racial oppression, which is not to suggest equivalence. Thus, it has examined how class is played out in the lives of blacks within the larger white-dominated society and within the black community. Globally, black Marxism examines how African and diasporan societies where blacks constitute the majority, such as the Caribbean, are affected by Western-dominated capitalism. It also examines class dynamics among blacks within those societies. That same tradition of thought is the foundation of the class analysis paradigm here.

for the mechanics of exploitation and oppression but also a theory of liberation through social revolution—something purely race-based, antiwhite-based, and purely class-based theories had failed to do. Racism initially was a by-product of economic competition between groups for wealth and power, but over time it was formally enlisted and remains vital to economic domination. Also over time, citizens of a racist society, both the bigots and quite often the targets of bigotry, come to subscribe to racist tropes.

Global black liberation requires a dialectical theory, method, and practice to end both white supremacy and economic oppression. In other words, Africana scholars and activists of the black Marxist paradigm advocate the synthesis of Pan-Africanism and Marxism as absolutely essential to achieving global black liberation. According to the class analysis paradigm, historical materialism is necessary but not sufficient for explaining the global black experience. Historical materialism starts with the fundamental material need for preservation. Humans organize themselves, most often nondemocratically, into divisions of labor to use increasingly sophisticated productive forces or tools that allow them to interact with nature so as to meet the group's economic needs. In the process, surpluses of material goods are generated that are controlled by powerful minority classes that own the means of production. Thus, the larger, poorer, and politically weaker working classes fail to get their share of valuable social rewards.

As Vladimir Lenin defined them, "classes are groups of people one of which can appropriate the labour of another owing to the different places they occupy in the definite system of social economy" (1977: 421). With each mode of production there exists a dialectic between the haves and the have-nots, who struggle for dominance. The contradiction and struggle are based on the social nature of production and the various private forms of appropriation. Although both are transmodal (e.g., slave system, feudalism, capitalism), in character of contradiction and struggle they are unique. Throughout the time period of each mode, the dialectic between social production and private appropriation become sharper and more intense until they force a progressive change to another. For example, by the time some Pan-Africanists had adopted Marxism early in the twentieth century, they were seeking to aid the Pan-Africanist mission by helping end capitalism and its super-exploitation of black labor. They had come to appreciate that this super-exploitation and the robbing of black humanity and dignity were intrinsic to a capitalist society.

White Marxists made the case that the black struggle for liberation would fail until capitalism itself was dismantled. In their view, the political agents of the state, on behalf of capitalists, had no material interest in terminating the exploitation of black labor and the humiliation of blacks as a people, and certainly blacks could never dismantle capitalism on their own. In fact, according to Glenda Elizabeth Gilmore, "The capitalists used 'state power' to 'cow [black] workers and keep them in constant subjection by the daily threats of conviction and imprisonment for alleged offenses.' The rule of thumb among the police was 'When in doubt, arrest a Negro'" (2008: 93). Not only were blacks prevented from effectively fighting for their liberation on their own, but white communists in the United States as well as labor throughout the world wanted a social revolution where humanity would progress beyond capitalism to a more equitable mode of production and social order. This idea appealed to Richard Wright, a major Pan-Africanist, of whom Cedric Robinson wrote, "Marxist propaganda suggested to him that Blacks need not be alone in their struggle for liberation and dignity. The specter of a world proletariat, united and strong, Black and white, fascinated Wright" (2000: 294).

Nevertheless, the type of black-white union relished by labor and theorized by Marxists was far easier to visualize than to actualize. Capitalists had a centuries-long stake in stoking black-white racial strife, dating back to Bacon's Rebellion in the seventeenth century. The white working class saw only whites in control of industrial societies throughout the world, and this ethnocentric observation was enlarged by the long-held belief that Europeans had always ruled and that Africans had never built any complex civilizations on their own. While many senior leaders of communist organizations such as the American Communist Party tried enlisting blacks to hold equally senior-level positions, many rank-and-file white Marxists simply did not consider black comrades their intellectual equals, an idea that originated in the South: "Everywhere white Southerners looked, they saw black Southerners behaving according to white supremacy's dictates, and they took that behavior as an indication of black people's inferiority" (Gilmore 2008: 16).

Furthermore, white Marxists had a low tolerance for black nationalism, which they saw as conservative and a wasteful distraction. Among the early black Marxists to encounter this prejudice were Cyril Briggs, founder of the African Blood Brotherhood (ABB) for Liberation and Redemption in 1919; Lovette Fort-Whiteman, the first American-born black communist leader (according to Gilmore 2008); and Harry Haywood, a leading Marxist thinker on the African American national question. Briggs, Fort-Whitman, and Haywood failed to unite black liberation with Marxism, just as many Pan-Africanists would fail to do throughout the remainder of the twentieth century. Even so, despite the intellectual and political challenges, Briggs, Fort-Whitman, and Haywood saw the synthesis of Marxism and black liberation as indispensable. Made up to a large extent of veterans of World War I, ABB members were determined upon their return home to end the wanton white-led violence in black communities that had been occurring with impunity. The ABB started out as a revolutionary nationalist organization and served as a military vanguard. However, the American Communist Party eventually controlled it as a device to penetrate and propagandize the black community on its behalf. After Briggs and other ABB members, including Harry Haywood, proved to be indomitable, the party expelled them.

Again, an attempt to synthesize Pan-Africanism and Marxism had failed because of resistance by both blacks and whites, although for different reasons, but efforts continued throughout the twentieth century despite the obstacles. Robinson (2000) chronicled many of the clashes between Marxism and black liberation, whereas Robin Kelley (1994, 1990) and Gilmore (2008) saw and reported the infusion of black traditions in Marxist organizations and that of Marxism in black revolutionary organizations. Nelson Peery explains, in *Black Radical: The Education of an American Revolutionary* (2007), how he sought to advance internal colonial liberation for blacks in America after having fought against fascism abroad in World War II. To reach that goal, many Pan-Africanists knew that a materialist conception of history and organized revolutionaries were required to dismantle capitalism.

Eventually, "just as Malcolm X moved to a position of revolutionary nationalism . . . the Student [Nonviolent] Coordinating Committee (SNCC), the League of Revolutionary Black Workers, the Congress of Afrikan Peoples, and the Black Panther Party would move from black nationalism to . . . black Marxism" (Dawson 2001: 199).

The synthesizing of Pan-Africanism and Marxism by activists such as Lovett Fort-Whiteman, Cyril Briggs, George Padmore, Walter Rodney, Amilcar Cabral, Kwame Nkrumah, Claudia Jones, Assata Shakur, Amiri Baraka, Abdul Alkalimat, and many others became known as the black radical tradition. Synthesized Pan-Africanism and Marxism, notwithstanding the individual limitations of each, offered the best chance for black liberation through social and self-transformation (Rabaka 2009). The black radical tradition clearly represents a distinct conception of being and, consequently, of revolution—one that is not based on the Marxist construction of the proletariat but rather takes liberation as its vision—and its subject is the global black experience ("The Black Radical Tradition," n.d.). The dialectic of proletariat and bourgeoisie was deemed inadequate by black radicals and replaced with a dialectic of imperialism and liberation. Black Marxism is largely reimagined as an indispensable part of the black radical tradition, which is rooted in the uniquely historical, cultural, and political perspectives of the black experience.

Not surprisingly, black Marxist theorists resolutely claim that Marxism not only remains relevant in the twenty-first century but is as relevant today as it ever was, given the growing inequality and fragility of global capitalism. As Terry Eagleton puts it in *Why Marx Was Right*, "You can tell that the capitalist system is in trouble when people start talking about capitalism" (2011: xi). Capitalists and the state apparatus that they dominate are unable to stop growing inequality from threatening the capitalist economy, a crisis on which the Occupy movement and other anti-oppression movements seek to center public attention. The top 1 percent are continuously appropriating a greater share of economic growth. The shares of the economic expansion that went to it during the Clinton and Bush postrecession years were 45 percent and 65 percent, respectively, but by 2010, after "liberal state actors" had controlled the Obama White House and the Senate, it appropriated $288 billion, or 93 percent, of postrecession growth. The top 0.01 percent, a mere fifteen thousand families, received over $114 billion, or 37 percent, of the growth (Rattner 2012).

One of the reasons that global capitalism's center is failing to hold is the potentially threatening impact of this new age of creative destruction. Non-Marxist economists Erik Brynjolfsson and Andrew McAfee, in *Race Against the Machine: How the Digital Revolution Is Accelerating Innovation, Driving Productivity, and Irreversibly Transforming Employment and the Economy* (2014) and *The Second Machine Age: Work, Progress, and Prosperity in a Time of Brilliant Technologies* (2012), and Martin Ford, in *The Lights in the Tunnel: Automation, Accelerating Technology and the Economy of the Future* (2009), all discuss how creative destruction has rewarded capitalists and most Western citizens and consumers—particularly during the Fordist era—but is now threatening capitalism.

Creative destruction impelled capitalist enterprises to continuously increase productivity and innovation in pursuit of greater profits. Throughout the industrial and early postindustrial periods, while such advancements were occurring, the amount of labor per unit production was decreasing in the aggregate, but more labor was needed because of growing markets. The new production technologies of today and those to come in the next era that Brynjolfsson, McAfee, Ford, and others describe point to a situation where innovation and productivity are increasing at a pace so fast that markets cannot possibly grow quickly enough to absorb all new products. Thus, overproduction continues, and more and more credit is used by the working class to fund its consumption. That credit becomes increasingly tenuous as workers' roles in production become more marginal and redundant with the progressive transformation of productive forces.

Creative destruction is threatening global capitalism not only technologically but also financially. In other words, new financial products as instruments of speculation are also fueling unsustainable growth in inequality. With sustained slow job and wage growth and with household income stagnating in some areas and declining in others, workers are left with insufficient purchasing power; Wall Street and the global investor class that it represents hire brilliant talent to create and operate financial instruments such as derivatives and credit default swaps, which even Warren Buffett sees as time bombs and "financial weapons of mass destruction" (Buffett 2003: 15). For example, it has been reported that a quarter of Harvard undergraduates fifteen years after graduation are in banking and finance. Also reported is that operations research and financial engineering have become the most popular undergraduate majors at Princeton's School of Engineering and Applied Science (Johnson and Kwak 2010)—not civil, electrical, or mechanical engineering, fields in which students are trained to build value-added products for the "real economy."

Afrocentrism

Between the late 1980s and early 1990s, and for a variety of reasons, black Marxism as an intellectual movement collapsed. Probably the most important cause was the demise of the former Soviet Union along with the strident reign of Reagan/Thatcher neoliberal economics. In the wake of this symbolic "end of history" event (Fukuyama 1992), black Marxists were perceived as deflated, on the defensive, and, falsely, a spent force. As a result, they lost—at least temporarily—black intellectual support. In the wake of the collapse of black Marxism and other revolutionary intellectual movements, Afrocentrism ascended as the beneficiary. That said, black Marxism played an instrumental role in the Black Radical Congress of the late 1990s and early 2000s.

Afrocentrism is the interpretation of, analysis of, and engagement with the life experiences of peoples of African descent—from their perspective and in their interest. According to Molefi Kete Asante, the Afrocentrist paradigm's most prolific exponent, "The Afrocentric enterprise is framed by cosmological, epistemological, axiological, and aesthetic issues" (1996: 256). As a result, it imputes a universal core to the black experience with respect to belief system, philosophical orientation, and cultural tastes and practices.

Many critics of Afrocentrism do not believe there is a common framework for how black people see the world or trust that they could have collective interests. This is curious, since everyone acknowledges the existence of Eurocentrism, which sees the world through the eyes of whites and engages it through white experiences and interests, and nearly everyone—at least those not of European origin—wants to see it dramatically held back, if not eliminated. Unlike with Afrocentrism, no one questions the existence of Eurocentrism, presumably because it has a global infrastructure that is replete with (1) dominating states capable of projecting power around the globe; (2) the world's most esteemed universities, which primarily advance Eurocentric knowledge; (3) many of the most powerful transnational corporations, which create and satisfy markets; and (4) media conglomerates that shape what Eurocentrists see as commonsense, accepted perspectives about social groups and social life—again from the standpoint of and in the interests of Europeans. This global weblike infrastructure, or hegemonic superstructure, reinforces a historical Eurocentrism that named the planets of the galaxy, nearly all of the nations and peoples of the world, and the world's most powerful institutions.

Not questioning Eurocentrism while disparaging Afrocentrism suggests that a "culture-centrism," with its focus on a people, exists only if it is hegemonic. Eurocentrism is hegemonic in a culture-centrist way, and that hegemony is reinforced by institutions that have dominated across space and time—at least for the past few hundred years. Antonio Gramsci's (1975) notion of hegemony is a domination so complete that the dominated consent to the routines and proscriptions designed by and in the interest of the dominating group. This has been manifestly obvious in recent centuries of African history.

Clovis Semmes defines cultural hegemony as "the systematic negation of one culture by another" (1992: 1), which is a fitting description of Europe's cultural hegemony over Africans. Centuries-long European imperialism fostered Eurocentrism and was itself manifested as an elongated *maafa* (ethnocide) of enslavement, colonialism, Jim Crow, apartheid, and other forms of oppression. Amilcar Cabral says that Afrocentrism is but one attempt to create a "theoretical weapon" (1973) to challenge Eurocentrism in Africa and the African diaspora. In other words, in its modern form Afrocentrism is epiphenomenal to Eurocentrism after the fifteenth century.

Michael Dawson, in *Behind the Mule: Race and Class in African-American Politics* (1994) and *Black Visions: The Roots of Contemporary African-American Political Ideologies* (2001), fundamentally affirms Cabral's view of Afrocentrism without calling it such; rather, he calls his version "linked fate": "A construct of linked fate is needed to measure the degree to which African Americans believe that their own self-interests are linked to the interests of the race" (1994: 77). For Dawson, the black experience of European imperialism has been so horrific that the outcome is a linked fate among blacks. That is, African Americans and others of African descent express strong notions of linked fate with members of their racial group. Moreover, Dawson claims that a "black utility heuristic" exists among individuals to help them identify and pursue common goals (1994: 57). Such a notion is in fact Afrocentrism without the label. Similarly, Pan-Africanists believe that peoples of African descent around the world and over the past five hundred years have shared their liberation struggles against European imperialism.

Neither Dawson's narrow though important linked fate nor the broader ideals of Pan-Africanism are considered controversial. In fact, they pass for prima facie cases. To understand why Afrocentricism by comparison is much more controversial, one should read James Stewart's article "Reaching for Higher Ground: Toward an Understanding of Black/Africana Studies" in the *Afrocentric Scholar*, in which he says:

> Two distinct claims are generally identified with the concept of Afrocentricity. What will be termed the "strong claim" is the assertion that the liberation of peoples of African descent requires a psychological reorientation that focuses on reconstructing selected aspects of traditional African psychology, values and behaviors in the present. The "weak claim" entails the position that liberation requires that top priority be assigned to the interests of peoples of African descent in social and political intercourse with other collectives. . . . Most of the criticisms directed at Afrocentricity have relevance only with respect to the strong claim. (Norment 2007: 428)

The strong claim is founded on one of the more salient points in Afrocentrism, as expressed by Cheikh Anta Diop: "For us, the return to Egypt in all fields is the necessary condition to build a body of modern human sciences, and renew African culture. . . . Egypt will play, in a rethought and renewed African culture, the same role that ancient Greco-Latin civilizations play in Western culture" (1974: 411). An Afrocentric perspective today overlaps with an idealized perspective that is traceable to ancient Egypt or Kemet and other classical African civilizations. The ontological and epistemological assumptions are fundamentally the same today as they were over five millennia ago. Although circumstances have changed profoundly over that time, the "strong claim" Afrocentric view is that the philosophical and intellectual roots from Kemet have provided African peoples with an outlook that helps them overcome whatever political, economic, and social challenges they encounter.

Many critics are dubious about such a proposition. The embodiment of these philosophical and intellectual roots in Maulana Karenga's (1980) Kawaida theory includes Nguzo Saba, and in Asante's theory it includes Njia. Afrocentrism wants us to place African ideals at the center of any analysis of African culture and behavior. However, what Asante means by "African" is a composite African, not specifically a Yoruba, a Ndebele, or a Zulu. It is not totally clear who will construct the composite African and his or her composite ideals on which to build; nor is it clear how contemporary black people's perspective and behavior should be measured in relation to the formulated African ideals as well as the pursuit of strategies for raising their Afrocentric consciousness.

Another problem with the idea of a composite African is that it does not account for class differences: "Today, Afrocentricity as a theoretical model has failed to develop a class analysis which takes into account the significance of political and economic power differences and the problems which may arise within the African community as a result" (Akinyela 1995: 27). Frantz Fanon, in *Wretched of the Earth*, cautions Africans and other peoples who have been colonized by Europeans (the wretched of the Earth) to be careful of their elites:

> Before independence, the leader generally embodies the aspirations of the people for independence, political liberty, and national dignity. But as soon as independence is declared, far from embodying in concrete form the needs of the people in what touches bread, land, and the restoration of the country to the sacred hands of the people, the leader will reveal his inner purpose: to become the general president of that company of profiteers impatient for their returns which constitutes the

> national bourgeoisie. . . . The people stagnate deplorably in unbearable poverty; slowly they awaken to the unutterable treason of their leaders. (1963: 166, 167)

Unfortunately, Fanon's warnings have been borne out in essentially all postcolonial African and Caribbean countries, from Ghana in 1957 to South Africa in 1994. For Afrocentrism to contribute to global black liberation, something it is certainly capable of doing, it is paramount that a class analysis be incorporated in its theoretical framework.

Class criticism aside, although it is hugely important, most of the criticism of Afrocentrism is directed at the "strong claim," while there is broad support for and subscription to the "weak claim." According to Makungu Akinyela, Asante's "argument . . . and his focus on making African people subjects of their own lived experience, rather than objects of European and Euro-American study, is relevant and necessary" (1995: 24). Perhaps Lerone Bennett (1984) was calling for support of Afrocentrism's weak claim when he discussed the need to develop a new, non-Eurocentric frame of reference that transcends the self-interest of white concepts of black people. Asante, Karenga, and other leading Afrocentric thinkers have called for Afrocentrism to be *the* paradigm of Africana studies. If it were primarily Stewart's "weak claim" form, many might agree that it should be paramount. In fact, one could argue that it is taken as given that the weak claim is the bedrock of Africana studies. As Martha Biondi sees it, "The articulation and defense of a 'Black perspective' defined the field [of Africana studies] from its inception" (2012: 245). Without this perspective, she claims, Africana studies would be no more than the study of black people, something that is recognized even by white intellectuals openly and indefatigably out to promote European imperialism.

Says Akinyela:

> Do these criticisms mean that Afrocentricity . . . should be abandoned? On the contrary, while some New [African] intellectuals have condemned Afrocentricity as irrelevant or reactionary . . . , Afrocentrism's strongest argument is in its call for a counter-hegemonic discourse to break the intellectual and moral legitimacy of the Eurocentric bourgeoisie on the minds and lives of

> the African, Asian, and Latin American world majority. I would add that it is strategically necessary to total human liberation to also break the hegemonic domination on the minds and lives of working men and women of European descent who are also in [the] cultural and political control of the ruling classes. The Afrocentric claim, that African people must construct a new African identity and must begin to perceive and interpret the world in its entirety from an African psychological, spiritual, and cultural frame of reference, is a correct one. (1995: 31)

Africana Womanism/Black Feminism

As a body of literature and a paradigm of Africana studies, black feminism is at least a generation older than Africana womanism, which began to command significant scholarly attention only in the early 1990s, when Afrocentrism was in its crescendo. As early as the 1970s, however, black feminism had emerged, on the one side, as a reaction to the androcentric orientation of black liberation organizations and literature, such as in the Black Panther Party and the *Journal of Negro History*, and, on the other side, as a reaction to the white affluent women-centric feminist movement, which articulated its orientation in, for example, the National Organization for Women and *Ms.* magazine. Audre Lorde (1984) expressed the exasperation many black women felt about their expected responsibility of educating black men and white women about black women's greater exploitation vis-à-vis other demographic groups. Black women wanted a discourse that would promote understanding among themselves and with sympathetic outsiders about being oppressed people of color, poor, and female. With the founding of women's studies in universities around the country in the 1970s, the 1980s, and beyond, black women engaged in hand-to-hand intellectual combat for space and credibility in the academy and in mainstream publishing.

At the same time there began a trickle of important scholarly products focusing on black women. In 1970 Toni Cade Bambara produced the first anthology about black women, which was followed three years later by Gerda Lerner's 1973 history on the aspirations and struggles of black women in white-supremacist America. By the early 1980s, bell hooks had begun to produce black feminist books such as *Ain't I a Woman: Black Women and Feminism* (1982) and *Feminist Theory: From Margin to Center* (1984). By the mid-1980s sociological and historical works as well as much literary criticism on black women were being regularly produced.

Long before this time, however, back in 1892, Anna Julia Cooper published what is now recognized as the founding black women's studies text, *The Voice of the South by a Woman of the South*. Cooper's scholar-activist orientation was similar to that in the work of her Pan-African friend W.E.B. Du Bois. Not surprisingly, American black women traced their intellectual and activist roots back even further, reverently and enthusiastically referencing the inspirational work of Harriet Tubman, Sojourner Truth, Ida B. Wells-Barnett, and Mary McLeod Bethune, among others (Guy-Sheftall 1984). Black women's studies sought to distinguish itself by resolutely refraining from a "great black women of history" approach. Rather than merely studying current and past "great black women" and waiting for another one to appear, black feminist scholars focused on the experiences of "ordinary" black women whose "unextraordinary" actions provide important insight for women seeking to overcome oppression in their daily lives (Guy-Sheftall 1984).

Black feminist thinkers such as Patricia Hill Collins "place black women's experiences and ideas at the center of analysis" (1991: xii). This black female centrality is a parallel to the Marxist placement of class oppression and struggle and to the Afrocentrist placement of African culture and knowledge at the center of analysis. In Marxism individuals' subjective positions depend on their material relationship to the means of production; in Afrocentrism black individuals reflect core African values and a core cosmological outlook. For Collins, "Black feminist thought encompasses theoretical interpretations of black women's reality by those who live it" (1991: 22). This involves a grounded theoretical approach involving careful study of countless black women, from various social backgrounds, struggling against a matrix of oppression and using their own viewpoint to make sense of and negate it. This approach affects in some instances other black women, both in their ideological standpoint and in their reactions to the matrix of domination, creating a spiraling pattern repeated ad infinitum and ultimately representing an imperfect, counter approach in which only great black women are studied.

Africana womanist intellectuals look askance at black feminism, because, as they see it, feminism's priority is to end sexism and the under-advancement of women that it causes. Feminism, as conceived by its affluent white founders, had only one -ism—sexism—to eliminate in order to be fulfilled as citizens of a hegemonic capitalist nation. Black women, however, most of whom are poor and working class, would have to fight against two other -isms simultaneously—racism and classism. As Paula Giddings (1984) notes, they worry more about racism as the root cause of their problems. Vivian Gordon points out that "the [feminist] movement fails to state clearly that the system is wrong; what it does communicate is that White women want to be a part of the system. They seek power, not change" (1987: 47).

Another criticism of black feminism, whether right or wrong, is that women tend to perceive all men as the enemy, just as most blacks at one time believed that—although not all whites mobilize capital and occupy positions of power as a means of racial oppression—all whites stand by and benefit from that oppression. In fact, most feminists acknowledge that historically not all men lived off questionably acquired property and had the institutional power to oppress women; however, as citizens of a patriarchal society, they stood by and reaped the benefits. In a black feminist or Africana womanist framework, according to Joyce Ladner, "Black women do not perceive their enemy to be Black men, but rather the enemy is considered to be the oppressive forces in the larger society, which subjugate *Black men, women and children*" (1971: 282; emphasis in original).

An interesting question is whether Toni Morrison's *Beloved* would be so beloved if Sethe had killed her daughter to keep her from suffering the injustices of sexism instead of the cruelties of slavery. Would it have the same power of wisdom and believability? Giddings (1984) suggests that it would not when she acknowledges blacks' rejection of white women's comparisons of their status to that of blacks and their referring to themselves as "niggers." Black feminism and feminism in general sort out these issues. Among the more supple works on this dialectic is that of Collins, who states, "Rather than emphasizing how a black women's standpoint and its accompanying epistemology are different from those in afrocentric and feminist analyses, I use black women's experience to examine points of contact between the two" (1991: 207).

Black feminism is currently more developed, with a larger, more structured body of literature than that of Africana womanism—by far. One reason is simply that black feminism has been institutionally supported for more than two decades longer. There is obviously some overlap between black feminism and Africana womanism, but there are notable differences, too. One is that feminists and Africana womanists can have separate agendas. According to some Africana womanist writers, feminism is female centered while Africana womanism is family centered. Yet some black feminists assert that their fight for pay equity, affordable child care, and universal health care is not just for themselves but for their black families.

Clenora Hudson-Weems coined the phrase "Africana womanism" back in 1987, when the Afrocentric paradigm was in its ascendency, marked by the establishment of Africana studies' first doctoral program in which the afrocentric paradigm would predominate. She says:

> The first part of the coinage, *Africana*, identifies [the] ethnicity of the woman being considered, and this reference to her ethnicity, establishing her cultural identity, relates directly to her ancestry and land base—Africa. The second part, . . . *Womanism*, recalls Sojourner Truth's powerful impromptu speech "Ain't I A Woman?," one in which she battles with the dominant alienating forces in her life as a struggling Africana woman, questioning the accepted idea of womanhood." (1995: 22–23)

Hudson-Weems makes the additional claim for a distinct intellectual turf for black women:

> Neither an outgrowth nor an addendum to feminism, *Africana Womanism* is not Black feminism, African feminism, or Walker's womanism that some Africana women have come to embrace. *Africana Womanism* is an ideology created and designed for all women of African descent. It is grounded in African culture,

and therefore, it necessarily focuses on the unique experiences, struggles, needs, and desires of Africana women. (1995: 24)

Like Paula Giddings, some scholars of black women's experience ascribe more importance to racism in the matrix of domination of black women than they do to sexism and classism—the latter two appear to be of equal importance to bell hooks and others, whereas Africana womanist scholars like Hudson-Weems and Gordon arguably give racism primary importance. Then again, some scholars, rather than ranking the relative importance of racism, sexism, and classism, see greater utility in thinking of the three as circular or dialectical where the drive is for a richer understanding of how they interact. It would be interesting to see how scholars of black women's experiences view and rank racism, classism, and sexism and then to look at how randomly selected black women view and rank them. Such a comparison would better address whether the claim of both Africana womanism and black feminism for centralizing black women's positionality is merely an ideal or a true reflection of their lived experiences.

Among the black women scholar-activists who place more emphasis on class than on race or sex in the experience of black women are Angela Davis and Rose Brewer, who, while acknowledging racism and sexism, focus on how they serve classism, rather than how classism and racism serve sexism or how classism and sexism serve racism. In *Women, Race, and Class*, Davis (1981) criticizes the feminist movement for ensuring political progress for middle-class white women while providing mild rhetorical support for blacks and other colonized people. This work and others by Davis lay bare the ways in which much of feminism is not about ending oppression but rather about mapping paths to positions of power and privilege, primarily but not exclusively on behalf of middle-class white women.

Rose Brewer imagines a "radical Black feminism which centers patriarchy and capitalism in the context of racism" (2003: 113). She also builds on Maria Mies's (1986) patriarchal capitalism and what bell hooks (1984) calls racist patriarchal capitalism. In addition to bringing a sharper analysis of class, including the black radical class, to the study of black feminism, she shines a light on liberation, class struggle, and the entire black radical tradition and how they have been improved by black feminist interventions. Inequality and class distinctions are growing rapidly in the black community, as they are throughout America and around the globe. The question is "Will middle-class black women continue to value racial solidarity with their working-class sisters, especially those in poverty, or will they use their newly acquired positions to perpetuate inequalities of social class?" (Collins 1991: 65). Brewer and others provide some insight into how these class dynamics play in the black community and in the nation with respect to black women, the changing economy, and black radicalism.

Another contributor to black radical women's literature is Carole Boyce Davies, whose *Left of Karl Marx: The Political Life of Black Communist Claudia Jones* (2008) offers a biographical and analytical account of one of America's leading black communists. Davies explains Jones's monumental work and how she came to be buried—at the end of her much too short but highly productive life—to the immediate left of Karl Marx. The book is not a plug for Marxism; in fact, that Davies disapproves of its white male universality is clearly evident.

Radical Egalitarianism

Some Africana studies scholars do not conceive of the three grand paradigms as residing separately in thick granite siloes. Instead, they see them blended to form new hybrids (e.g., black feminism/black Marxism or Afrocentrism/Africana womanism) that are mostly used consistently but sometimes only situationally. In the future, we are also likely to witness more blended paradigms from Africana studies and environmental studies—for example, Afrocentrism/environmental justice, black feminism/ecofeminism, and black Marxism/eco-Marxism. This section provides a brief example of such usage beyond the three paradigms in Africana studies.

Africana scholars such as Manning Marable, Cornel West, and Tommie Shelby move within an Africana paradigm that can be called radical egalitarianism. They started with the fact that black societies everywhere share a centuries-long baseline of racial oppression and the fact that long before and long after that oppression got under way, black people shared a core of common cosmological, family, community, and other values. The shapers of radical egalitarianism adopted the "weak claim" form of Afrocentrism as formulated by Stewart (1992)—that is, the centrality of black experiences as seen from the perspective of black people.

From that point of departure, radical egalitarianists, while not seeing their works as black Marxist, see the opportunities for and the necessity of working beyond the black community writ large to eliminate racial oppression and thereby work toward global black liberation. Cornel West (1988) argues that there is a powerful and irreducible element of race that cannot be subsumed within class but acknowledges that black oppression and exploitation must place capitalism at the center of analysis. Manning Marable is more explicit, starkly asserting, "The race question is subsidiary to the class question in politics" (1995: 229). According to Michael Dawson, "Like West, Marable argues that the intersection of white supremacy and capitalist oppression has to be the starting point for understanding not only black oppression, but the condition of the disadvantaged in the late-twentieth-century United States" (2001: 223).

As for Tommie Shelby (2005), he argues that political blackness and "pragmatic nationalism" should replace racial and essentialized blackness: "Political race suggests that patterns which converge around race are often markers of systematic injustice that affect whites as well, and thus disclose how institutions need to be transformed more generally" (Guinier and Torres 2002: 20). Thus, a politically raced perspective looks at the ways that race camouflages unfair resource distribution among whites as well as blacks (Guinier and Torres 2002). Shelby's radical egalitarianism, like that of West and Marable, is not about either racial essentialism or race blindness: pragmatic nationalism is the view that "black solidarity is merely a contingent strategy for creating greater freedom and social equality for blacks, a pragmatic yet principled approach to achieving racial justice" (Shelby 2005: 10). For him, both political race and pragmatic nationalism are compatible with interracial cooperation.

In short, radical egalitarianism in Africana studies centralizes the black experience in class-analytic terms, leaving sexism as a conspicuous omission. The global black liberation project needs and so provides opportunities for alliances with nonblacks, which radical egalitarianism seeks to facilitate.

TABLE 2.3 AFRICANA STUDIES PARADIGMS

Paradigms	Units of analysis	Constructs	Exponents
Class analysis	Racial labor market segmentation	Working class; labor exploitation; labor redundancy	Cedric Robinson; Abdul Alkalimat; Robin Kelly; Rose Brewer
Afrocentrism	African ideals, culture, and knowledge systems	Composite African; Njia; Nguzo Saba	Molefi Asante; Marimba Ani; Maulana Karenga
Africana womanism/ black feminism	Black women's lived experiences and standpoint	Matrix of domination; culture of resistance; gender equity	Patricia Hill Collins; bell hooks; Vivian Gordon
Radical egalitarianism	Egalitarian dispensation and redistribution	Radicalism; equitable dispensation	Cornel West; Manning Marable; Tommie Shelby

Table 2.3 lists the four paradigms in Africana studies. It also lists the units of analysis, common constructs, and major exponents for each.

Table 2.4 lists the paradigms of both Africana and environmental studies. It is the starting point for a dialogic discourse, which is the first step in bridging the two fields and thus enabling them to address environmental ills in black and other communities of color and to enlist blacks in building sustainable futures.

Africana studies seeks the improvement of the black experience both locally and globally, but that cannot happen without the improvement of black people's environment, which requires black agency. Environmental studies seeks sustainable economies and communities. Since black communities are often the most degraded and least sustainable, its mission cannot be achieved without achievement of the Africana studies mission. This chapter discusses the integration of Africana and environmental studies objectives and perspectives as a way to complete the missions of both. In Chapter 3 I apply the multiple interdisciplinary perspectives of Africana and environmental studies to environmental challenges in black communities across and beyond the United States.

TABLE 2.4 LINKING AFRICANA AND ENVIRONMENTAL STUDIES PARADIGMS

Paradigms	Ecological modernization	Environmental justice	Eco-Marxism	Ecofeminism	Ecocentrism
Class analogy/black Marxism	Waste reduction; relations of production	Colocation of LULUs with poor/working-class communities	Racial labor market segmentation in polluting industries and territories	Class consciousness and power-over	N/A
Afrocentrism	Decentering Eurocentric approaches to environment-society nexus	Environmental racism; African ideals	Decentering Eurocentrism and environmental destruction	Decentering Eurocentrism and environmental destruction	N/A
Africana womanism/ black feminism	Interlocking systems of domination; androcentric technological fixes	Black women's lived experiences; LULUs	Capitalist-driven environmental destruction and black family health	Matrix of domination; logic of domination	N/A
Radical egalitarianism	Radical advancement of triple bottom line	Equitable distribution of environmental positives and negatives	Redistributive government; ending treadmill of production	Zero-sum politics; androcentric "mastery" of nature	Humans' intrinsic worth equal to that of other species

VIGNETTE 2.1. CYNTHIA HEWITT
Africana Studies at Morehouse College—African Ontology for a Green Future

KNOW THYSELF

—INSCRIPTION ABOVE THE ENTRANCE OF THE TEMPLE OF KARNAK, CA. 2000 B.C.

The development of an Africana Studies Institute at Morehouse College with a green and international focus has evolved from seeking cures to the sufferings of African people in Africa and throughout the diaspora and now includes self-empowerment to resist racist subordination. Providing assistance after the devastation of Haiti in 2003 by Hurricane Jeanne was the focus of "Positive Action" for that year's annual Africa Awareness Week. Morehouse students and faculty realized that Haiti (and other African and African diaspora societies) would remain vulnerable to such environmental devastation until its deforestation and its lack of sustaining use of plants were addressed. Following a talk by renowned ethnobotanist, Dr. Anthony Kweku Andoh, and cofounder of GENESIS (Growing Energy and Nutrition for Environmental Stability and Investing in Society) many of us became captivated by the fundamental fact that neither political movements nor economic systems but rather "knowledge of plants will set you free." A student reforestation brigade was formed, and with the help of former Ambassador Andrew Young and entertainer Wyclef Jean, students planted over three hundred trees in Haiti in 2004. With the shift in political forces (the ouster of Jean Aristide), our attention shifted to Ghana, where MPAGE (Morehouse Pan-African Global Experience) was started in 2009. Today, the MPAGE program forms the core of an international and green focus around which we are building the Morehouse Africana Studies Institute.

The leadership of Dr. Andoh helped ease the transition by introducing ethnobotany, or the study of culturally based uses of plants, the ultimate renewable resource, as a form of environmentalism, which is often perceived as insensitive to issues facing African people in the United States and Africa. The regard for new environmental dangers, such as rising seas, seems to coexist with continued disregard for people of color who are actually already hungry and displaced. However, one approach—what I call the back-to-the-land movement—which is oriented toward sustainable minimalist lifestyles, seems particularly compatible with Africana studies objectives. The movement is fueled by a prospect of dropping out of the carbon-intensive pollution system as a result of local sourcing of organically grown foods reminiscent of the "hippie," or more current, "hipster," orientation. How do we bridge this reality and the realities experienced by poor, often black and brown, people? I think the development of Africana studies can help bridge this gap by enhancing our knowledge of plant uses and by restoring memory, pride, and options to people of African heritage.

Africans traditionally lived in a reality that included both the past and the present and both material and immaterial dimensions. Their societies required important social roles involving careful learning and consistent practice related to the immaterial dimension—beings or forces, including identified ancestors, gods, and spirits. Many African people have an understanding of themselves as "humans," who represent only one among several organized forces operating and interacting in one or more of the multiple dimensions of reality. A green Africana studies program should thus not be only about Earth but about Earth and human beings in a multidimensional universe. A central understanding of African societies is that the limitations of perception, time, and speed, as explained by astronomers and physicists, make material existence only relative. That is, material reality is only one aspect of reality. All sorts of ingenious cultural adaptations of our material senses, appetites, and desires to the ultimate order of reality have been created by Africans for over 250,000 years. African knowledge systems also include complex understandings of plants and their uses, one such being the use of quinine to prevent and cure malaria.

African American history and society also evolved in parallel with those of the native people of the Americas, with whom blacks shared many traits, including a fundamental mother-centered culture (as existed, for instance, among the Cherokee and the Iroquois). Some ontological understandings and historic knowledge have been faithfully transmitted to us today, but so much of these social/philosophical conventions have been recklessly demolished with the coming of Western religions and market society.

Green Africana studies at Morehouse College means looking back and trying to reconnect and nourish indigenous knowledge so that we can potentially reseed the earth with social relations more green and sustainable. In America, it means looking to the black farmer and to the Depression-era grandmother who insisted on rinsing used bottles and folding aluminum foil for another use, to the irritating great aunt who offered bitter home remedies, to the uncle who got under the car and fixed it himself. The MPAGE approach seeks to connect American youth to African-heritage knowledge, especially in rural areas, and then to connect this knowledge to the "mother lode" in Africa. This cross-cultural dialogue is fundamental to the MPAGE.

The biggest challenge of renewing the wisdom of Africa is the lack of investment in traditional African education, including the oracles, the family, chieftaincy, and healing societies, among others. One cannot simply import piecemeal African elders and holders of knowledge and encase them in the antiseptic environment of Western universities. Different ways of learning have their own environmental prerequisites. I suggest that the end goal of green Africana studies must be to effectively connect the flow of resources back into the villages and towns that are still governed, if only in part, by ancient forms of wisdom and that still rely on using plants to meet their needs and then to allow that wisdom to manifest, grow, and adapt. We can then reconnect and learn from it, and African people in Africa and those of the diaspora can take their place within the back-to-the-land/grow local movement.

3/ Brownfields, Toxics Release Inventory Facilities, and the Black Community

Introduction and Background

Environmental policy, like any other public policy, is defined by the dominant social paradigm, which is the embodiment of the beliefs and values shaping our outlook on society, government, and individual responsibility. The dominant social paradigm in America is laissez-faire capitalism—a minimalist government on behalf of the average citizen—with continual economic growth and technology fixes for much of our personal and social ills (Smith 2012). It is also hegemonic and self-interestedly advocates a social stasis that solidifies soaring political and economic dispensation.

Fundamental change comes from those corners of society that are progressively against the dominant paradigm. Social change to end environmental problems is like most other social changes: the public pressures government to act, and the government in turn pressures corporations to change their business practices. Regardless of this political pressure, corporations loathe change, particularly regarding the environment and especially when their business models are working and meeting their bottom-line objectives. Pressure for economic and business change also can come directly from the public via the market, but the most significant changes in industry relating to the environment are mediated through the government.

Rachel Carson's 1962 *Silent Spring* and the environmental movement it helped spawn contributed to pressure on the federal government to control industrial air pollution through the Clean Air Act of 1963. The growing green power movement, along with other movements for broader progressive social change, in 1970 resulted not only in a more expansive Clean Air Act and the National Ambient Air Quality Standards (NAAQS) but also in the establishment of the Environmental Protection Agency. This was the supposed "command and control" era, when the government set minimum and rigid regulatory standards and the means of achieving them by industry, which had to be met if an industry sought to compete in a given market. In the 1980s, progressive movements, including green power movements, were a mere shadow of their former selves. This was the Reagan-Thatcher era, when the dominant social paradigm of laissez-faire capitalism was stridently hegemonic; it was also when command and control was replaced by market incentives to coax desired corporate behavior toward mutually beneficial outcomes for corporations and broader society.

But even strident neoliberalism could not withstand public demand for government action in the wake of the 1979 toxic waste scandal in the Niagara Falls neighborhood of Love Canal in New York. The following year, the Comprehensive Environmental Response, Compensation, and Liability Act (CERCLA) was initiated to clean up environmentally hazardous sites like Love Canal around the country, particularly in cases where the responsible party either could not be identified or was without insurance and the financial wherewithal to remediate the properties and the mental and physical health problems of affected people. There are 1,673 Superfund sites in the United States today.

The Toxics Release Inventory (TRI) Program came into existence as a result of public pressure for action against industrial negligence and abuse. For example, in 1984 the Union Carbide industrial disaster in Bhopal, India, killed some twenty thousand and injured some six hundred thousand. Around the same time, smaller accidents with the potential for a Bhopal-scale disaster occurred at a Union Carbide plant in West Virginia. The American public suddenly realized its vulnerability to deadly chemical releases from nearby industrial facilities about which they knew little, and once again public pressure resulted in government regulations that affected chemical companies. For example, the Emergency Planning and Community Right-to-Know Act (EPCRA) of 1986 authorized the EPA to report to the public, through TRI, about the acquisition, storage, development, processing, transfer, and release of criteria chemicals (the more toxic chemicals as identified by the EPA) at stationary facilities throughout the country.

Public pressure on government ultimately caused chemical companies to alter environmentally harmful and health-vitiating practices through a mixture of punitive minimum standards and market incentives, and it has resulted in some environmental improvements, albeit insufficient in light of the challenges. For instance, twenty years after industrial firms were required to disclose their toxics release inventory data, toxic releases were down over 60 percent despite a significantly larger industrialized economy. Thus, air quality in America is substantially better today because of improvements in stationary (i.e., industrial) as well as mobile sources. For instance, cars today are 90 percent cleaner than their 1970s counterparts (Smith 2012). The ecological modernization paradigm would say that such outcomes bear out its claims that we need only "technology-fix" our way out of industry-induced environmental problems rather than entertain economic approaches beyond capitalism as we have known it.

As established in Chapter 1, although Africana studies has not been completely missing in action in efforts to improve the environment in black communities, it has contributed only an infinitesimal share of its enormous intellect and energy toward fixing such communities' existing environmental ills. Africana studies has not prioritized the physical environment in black communities or engaged sustainability, and the environment is poorer as a result. As observed by Janet Currie (2011), black children in America are far more likely to be exposed to pollution in utero than white children, which increases their chances of poor health at birth and thus poorer outcomes throughout life, including educational attainment, professional employment, and stable families. One of many environmental threats to black children is lead poisoning, which can cause brain damage and thus behavioral problems. A recent statistic shows that African American children have incidence rates of lead poisoning five times those of white children (Bae 2012).

This chapter examines brownfields and major toxin-releasing facilities in black communities. Both categories of locally unwanted land uses (LULUs) are discussed in terms of their impact on black communities, what is being done to address them, and what Africana studies can do to address the ills they engender. African Americans face greater environmental problems because of their disproportionate proximity to environmentally hazardous sites. "'Risks' arise at the intersection of vulnerability and susceptibility compounded by a lack of information and inadequate access to health care. 'Vulnerability' refers to heightened opportunities for hazardous exposure. 'Susceptibility' refers to intrinsic individual factors that render some people more likely to get ill from exposure" (Burger and Gochfeld 2011: S65). With regard to the physical environment in black communities, Africana studies is uniquely qualified to create new knowledge, elevate awareness, and contribute to community efficacy. Such engagement starts with the greening of Africana studies.

Brownfields

According to the EPA, "A brownfield is a property, the expansion, redevelopment, or reuse of which may be complicated by the presence or potential presence of a hazardous substance, pollutant, or contaminant" (U.S. Environmental Protection Agency 2012). In other words, brownfields are sites of former commerce—from a gas station to a hulking industrial behemoth—that are either completely or partially abandoned. Although the commercial enterprise may be gone, the environmental hazards, or the perception of them, continue to threaten and frustrate safety and progress in the local community. There are between five hundred thousand and one million brownfields in the United States, which will cost between $700 billion and $1.4 trillion to remediate. Figure 3.1 shows an example of a brownfield. Remediating brownfields and redeveloping them for residential living, commercial activity, or green space are crucial to revitalizing the urban core. In addition to creating healthier communities that are free of blight as a result of massive brownfield redevelopment, we create employment, since the jobs required to redevelop the nation's brownfields cannot be offshored and do not require most workers to have advanced degrees.

Figure 3.1 Brownfield in a low-income black neighborhood in Toledo, Ohio

Much of the funding required to remediate America's brownfields has to come from the federal government, as states and municipalities simply do not have this kind of money and are obligated constitutionally to keep their books balanced. The requisite public spending for massive remediation is unlikely to happen, however, as long as Washington is influenced primarily by transnational capital, which sees this money as a waste because much of its profits are derived from global investments.

During the first two-thirds of the twentieth century, major capitalists such as Carnegie, Rockefeller, Vanderbilt, and Morgan, who derived much of their wealth from building and expanding America's infrastructure, advocated massive public investment. However, in more recent decades, Wall Street has shown substantially less interest in domestic industrial investment (Patterson 2013b). And as we have seen even in the wake of the financial collapse of 2008–2009, Wall Street typically gets its way, and its desired policies are enacted. Rather than massive public spending on brownfield remediation, it champions tax cuts that favor the top 1 percent so that, in part, more funds can be invested globally in search of the greatest return. One of the best ways for transnational capital to block extensive brownfield redevelopment is to castigate it as liberal big government run amuck with yet another welfare program. Such a racially tinged, stinging caricature is effective at turning many whites against public infrastructure policy, from which, ironically, they would largely benefit (Edsall 2006; Patterson 2013b).

Capital was becoming transnational by the 1970s, which is when brownfields began to proliferate. "Deindustrialization" is the term of choice for the outcome of transnationally oriented U.S.-based manufacturers choosing to close facilities in the urban core and move them to new locations, often in the U.S. South or the Global South, where wages, working conditions, and environmental regulations are lower (Wilson 1996). Lani Guinier and Gerald Torres (2002), using the metaphor of canaries in mines, describe what happens to blacks in America as a portent of what is likely to happen to middle- and working-class whites. Blacks have long been the first to suffer from a lack of health insurance coverage, reliance on check-cashing stores, structural and long-term unemployment and underemployment, and deindustrialization-influenced out-of-wedlock births as a result of a sharp and sustained rise in men in the community not being gainfully employed. Now, however, whites in increasing numbers are facing all of these markers of downward social mobility and family reconfiguration as a new normal. Deindustrialization, it seems, is like a noxious mine gas, but now it no longer primarily affects the canaries—people of color and the poor and uneducated; rather, all Americans who work for a living are affected by twenty-first-century deindustrialization and transnational capital. Initially, in other words, jobs left black communities and blacks suffered, but now jobs are leaving America and whites are suffering—in a way they have not experienced in decades. The metaphor of the black experience as the canary in the mine aptly shows what is in the future for white America.

Africana studies paradigms can be highly instrumental in an ongoing struggle to challenge the dominant social paradigm that tolerates blighted communities, inferior and rotting infrastructure, wasted human potential in inner cities, and contaminated soil and water in search of the most lucrative global investments. The paradigms of Afrocentrism, class analysis, black Marxism, black feminism/Africana womanism, and radical egalitarianism question some of the presuppositions of the more progressive environmental studies paradigms, produce more relevant and incisive studies, and mobilize a broader public to pressure government for more progressive outcomes.

Part of sustainable development is land recycling, including brownfield redevelopment, to avoid continual development of greenfields (i.e., virgin land). Typical brownfield contaminants are low concentrations of toxic materials such as heavy metals, arsenic, polychlorinated biphenols (PCBs), and hydrocarbons. Although brownfields should be remediated for environmental and public health reasons, the unspoken yet prevailing wisdom is that they have only low concentrations of hazardous materials that are unlikely to trigger acute public health emergencies. Local and state officials believe that if brownfields were a major imminent environmental and public health problem, they would be on the National Priorities List (NPL), which comprises the nation's 1,673 most dangerous hazardous sites.

The federal government is responsible for NPL sites, and under the polluter pays principle, the EPA holds responsible parties accountable for the environmental degradation. Since the public health threats posed by NPL hazardous sites are regarded as imminent and potentially severe, the federal government acts as the public's comprehensive insurance policy. That is, if they have the wherewithal, violators pay for remediating sites through CERCLA; if they do not or no longer exist, the government pays.

Brownfields were at one time under CERCLA control, which allegedly frightened would-be investors. Today, however, they largely fall under the purview of state and local governments, and the tools of remediation are increasingly less financial and "command and control" and more collaborative. Often, state and local governments do little more than dispense tax abatements, liability relief, and matching funds. In other words, rightly or wrongly, brownfield remediation today is less about environmental health and more about economic development. Local governments primarily try to improve the local tax base rather than privilege risk reduction from brownfields (De Sousa 2005). The regulatory relief that state and local officials offer to investors are uncapped liabilities stemming from uncertainties surrounding the hazards. According to Hunter Bacot and Cindy O'Dell:

> Recent state legislative efforts generally "relieve developers of uncertain liability risks and otherwise support regeneration efforts. . . ." Whereas states oversee general program direction through legislation and therefore wield regulatory authority for brownfield policies, local governments generally direct and manage the mitigation and redevelopment of specific brownfield properties. Local government departments of planning and economic development typically direct local brownfield policies and programs. (2006: 144)

In 2002 the federal Brownfield law amended CERCLA through the Small Business Liability Relief and Brownfields Revitalization Act, which was crafted to spur brownfield redevelopment by (1) encouraging small businesses to partake and (2) preventing liabilities from being levied on "innocent landowners" who acquire brownfield sites. The idea is that, if investors know up front the limits of their risk and know that they will not be exposed to unpredictable liabilities for previous contamination, more will back brownfield projects. As Flannary Collins observes:

> While the changes in the Act seem to focus adequately on the economic and environmental justice issues involved with brownfield redevelopment, the changes may have come at the expense of environmental concerns. Specifically, the Act's limitation of liability of certain purchasers of brownfield property is aimed at spurring development, increasing jobs, and encouraging the redevelopment of brownfields. However, this release from liability may have negative impact on environmental cleanup because the government will be burdened with much of the liability and subsequent cost of cleanup. (2003: 305)

Some critics of the brownfields revitalization act admit that CERCLA was too much of a cudgel against businesses, but they say it has swung too far in the opposite direction to support investors with too little focus on the environment and public health.

State and local governments are often reticent about brownfield sites because announcing them may frustrate efforts to attract two groups that would promote economic revitalization: first is the "creative class," as Richard Florida (2002) describes it; second is the well-known and much-heralded investor class. One worthwhile challenge is changing the public view of brownfields: from seeing them as a liability to seeing their potential to become assets once remediated. Brownfields are more economical than greenfields in that they offer "an existing infrastructure such as water and sewer distribution and collection networks, roads, and power supply," and once the are remediated, that full potential can be realized (Nagengast, Hendrickson, and Lange 2011: 298). Amy Nagengast and colleagues looked at energy consumption and greenhouse gas emissions resulting from traveling to greenfields compared to traveling to brownfields in six sample cities. Greenfields were found to be six times further from the city center. "Brownfield commuters had on average 37% lower energy and 36% lower greenhouse gas emissions for their commuting trips" (2011: 303).

Trying to sell brownfield redevelopment to investors before remediation (i.e., prioritizing the needs of investors over the needs of environmentally diminished neighborhoods) is frustrating first and foremost to the residents of neighborhoods where brownfields are located but also to environmentalists. Properties where the worth of post-redevelopment is expected to be less than the cost of remediation are not likely to be remediated. Often the brownfields with the lowest market value have the greatest contamination and are located in the least desirable neighborhoods. These "upside-down" brownfields, as Linda McCarthy (2009), calls them, are at the bottom of the list and are the least likely to be redeveloped according to the dominant social paradigm of neoliberalism. The blacker the census tracts, the greater the likelihood that brownfields exist, the greater their prospects of being upside down, and the greater their chances of not being redeveloped.

Upside-down brownfields are disproportionately located in black neighborhoods. According to Randy Stoecker, "[The] community's tendency is to preserve neighborhood space as a use value for the service of community members, while capital's tendency is to convert neighborhood space into exchange values that can be speculated on for a profit" (1997: 5). Africana programs and community development corporations (CDCs) can work collaboratively and thus better engage with black neighborhoods that often contain some of the most hazardous brownfields.

Many Africana scholars write off CDCs as tools of the establishment rather than seeing them as transformational change agents in the black community. Some of this skepticism stems from the tension between community organizing, which is more the founding tradition of Africana studies, and community development, which is more in line with market-oriented solutions that have come to characterize CDCs. The sharpest criticism is that CDCs are based in the community but not necessarily community-based. However, we probably should not take this criticism too far, since universities offering Africana programs are also *in* or *near* the community, although often not *of* it. Furthermore, universities uphold the political and economic status quo and so privilege market-based solutions far more than they do political and economic redistribution. Despite this fact, Africana scholars and practitioners do work with and in universities. In fact, the prevailing wisdom is that those in the field—notwithstanding ivory tower, market-oriented activities—can have a transformative impact on the black community through university-based engagement. This invites the question: How can one in principle refuse to work with CDCs because of their tilt toward market-based solutions but work in and with universities despite not only their tilt toward but their undergirding of market-based solutions to social problems? There is very little ground on which to defend that position.

Thousands of CDCs across the country seek to improve housing stock and other physical structures, civil society, and economic institutions in impoverished and working-class communities, many of which would otherwise see almost no capital infusion. Although even the most successful CDCs are insufficient, when they do succeed in attracting funding, there is at least the prospect of economic development in low-income communities. CDC board members are usually community leaders and professionals who use the skills and social network to pursue political support for federal, state, and local funding.

Many Africana programs provide, if not require, courses in research methods applied to urban black communities. Through such courses, among others, students obtain internships and participate in community-based research with CDCs in black communities in cities where Africana studies programs are securely established. They gain valuable experience that they can use after graduation. CDCs also can act as structured social laboratories for relevant and publishable research.

L. W. Green and colleagues define community-based participatory research (CBPR) as "systematic inquiry, with the participation of those affected by the issue being studied, for the purposes of education and taking action or effecting change" (quoted in Gonzalez et al. 2011: S166). Table 3.1 summarizes the dimensions and metrics of

TABLE 3.1 MEASUREMENT INSTRUMENTS AND METRICS TO ASSESS COMMUNITY CAPACITY

Dimension	Metrics to assess
Leadership	Number of individuals playing various leadership roles
Participation	Counts of individuals participating at various levels
Skills	Self- or external rating of relevant skills, including technical/scientific and organizing
Resources	Inventory of human, social, and financial resources
Social organization networks (community linkages)	Organizational networks
Sense of community	Feelings of connection, support, and collective problem solving
Community power (empowerment, perceived control)	Ability to influence decisions; partnerships with institutions; perceived impact on policies; perceived control at multiple levels
Communication	Content, frames, and scope of written and verbal communication with community initiative
Group cohesion	Sense of belonging to group; ownership of what group does
Community capacity	Historical narrative or cumulative scale of previous dimensions
Social capital	Mutual trust and reciprocity; strengthening neighborhood social networks; all becoming normal

Note: Adapted from Freudenberg, Pastor, and Israel 2011: S125.

green Africana community-based CBPR and engagement concerning the environment.

One cutting-edge area of research for green Africana studies is transforming brownfields into renewable energy sites known as brightfields—that is, land remediated for its next recycled use. Going directly from a brownfield to a residential community on the land is the most intensive and expensive remediation. Going directly to an industrial facility is the cheapest and least intensive. Remediating the contaminated soil and water to certain levels and then using the site for energy (e.g., wind, solar) collection and distribution is another less expensive option for brownfield development, for which the National Renewable Energy Laboratory (NREL) provides research and development funding. Figure 3.2 shows a solar field at the 180th Fighter Wing Air National Guard Base at the Toledo Express Airport, which is what a brightfield might look like. Because it is much better to employ brightfields research on the front end, Africana studies can work with local citizens on surveys and focus group discussions about the potential value of brightfields in their respective communities. Faculty and students can work with CDCs in black communities to uncover the potentially harmful effects of technologies such as solar and wind, and then, once the facts are known, survey residents' preferences concerning these technologies in their communities.

Many states and municipalities set goals for making renewable energy part of their future energy portfolios. The idea is to consistently

Figure 3.2 Solar field at the 180th Fighter Wing Air National Guard Base at Toledo Express Airport

replace fossil fuels with renewables. The state of Michigan's Renewable Portfolio Standard set a goal of 10 percent of its energy use to be derived from renewables by 2015. Soji Adelaja and colleagues (2010) note that this goal could have been met a long time ago, and exceeded by 50 percent, if all of Michigan's brownfields were converted to wind and solar energy production sites.

NREL supports excellence in research and development in renewable energy for the Department of Energy (DOE). One major area of focus is movement from brown to green—that is, from brownfields to renewable energy sites. An example is *Feasibility Study of Economics and Performance of Solar Photovoltaics at the Former St. Marks Refinery in St. Marks, Florida* (Lisell and Mosey 2010).

Similar to the DOE's efforts are the EPA's Re-Powering America's Land initiatives, which invest significant resources (although much more is warranted) to siting renewable energy sourcing (wind, solar, biomass, and geothermal) on contaminated land. The EPA recognizes that as much as twenty-two million acres of contaminated sites have potential for renewable energy production. Just one among many initiatives in this arena is converting former gasoline stations—now characterized as brownfields because of leaking underground storage tanks—into alternative-fuel stations. EPA and NREL maps show that many existing infrastructures would support transportation because their locations are near considerable automobile traffic. Irrespective of the renewable source, vehicles will have to be repowered away from home (Johnson, Hettinger, and Mosey 2011).

Yet another brown-to-green activity is the conversion of brownfields into urban agricultural gardens. The Africana studies program at the University of Toledo (UT) worked with the Toledo Community Development Corporation (TCDC) to redevelop the numerous brownfields in its service area, which is in the heart of Toledo's black community. One of these brownfields was remediated and has been redeveloped as the Fernwood Growing Center (FGC). TCDC's service area census tracts have a poverty rate three times the national average, a median household income half the national average, and a college graduation rate only one-fifth the national average.

Funding for the TCDC remediation and redevelopment initiative, with the UT Africana program's help, came in the form of a $200,000

Brownfield Cleanup grant from the EPA and another $50,000 from a Toledo Port Authority Pre-development grant and the Ohio Community Development Finance Fund. These moneys were used for the environmental remediation of a two-acre donated brownfield site and for generating a business plan for its redevelopment. The plan is for FGC to house four large hoop houses with an aggregate growing space larger than a football field. An additional plan is for hydroponic farming of tilapia and raised beds for an assortment of greens and herbs.

Before starting this project, the TCDC studied and mapped food-purchasing patterns in its service community, particularly regarding fresh produce. Residents reported consuming beans, potatoes, cabbage, and other types of greens and said that they would consume much more if they did not have to travel more than two miles out of their food desert (the geographical space where healthy foods are inaccessible and unaffordable) to purchase them and if they were less expensive. Price and distance are major issues because more than one in three households are in poverty in the service area census tracts, and nearly one in five does not own a vehicle. The study also surveyed restaurants and grocery stores around the city about their potential for purchasing from the FGC.

The vision of TCDC's executive director, E. Michelle Mickens, is for the FGC to employ approximately twenty-five community residents, increase consumption of fresh produce in the community, improve citizens' knowledge of healthy food preparation and consumption, and help them acquire a sense of stewardship of their environment. Many lives will be improved dramatically by the FGC. Consider that approximately two-thirds of families with children under five in the TCDC service area are impoverished. Now combine that statistic with the medical fact that the infant brain uses some 60 percent of total nutritional energy consumed and that it continues to grow rapidly until children are about the age of five. Undernourished small children with insufficient energy to fuel brain growth can suffer from stunted cognitive development, which manifests itself in impaired attention, comprehension, learning, memory, and so forth. All of this means that the stunting of human potential itself is quietly taking place in many impoverished families with small children.

Nutritional enhancement is one benefit of the FGC; another is social capital, which the FGC and other community gardens can generate as a form of positive reciprocity—"I do for you as you do for me"—that reverses the negative reciprocity of "I do to you what you do to me." In other words, social capital fosters solidarity. Social capital is particularly scarce in TCDC-type census tracts all over the United States—the very communities that need it the most. These communities suffer doubly because (1) they lack material resources to get ahead (e.g., jobs, quality education), which come primarily from outside the community, and (2) they lack the in-community social resources (e.g., networking, trust, community-based organizations) that would enable them to collectively amass those external material resources.

Finally, the FGC can provide classes for local residents with the help of UT's Africana studies program. These classes will introduce residents to methods for improving their nutrition, growing their own urban gardens, joining the urban agricultural movement, acquiring brownfield remediation skills, and reimagining and rebuilding urban communities that have been devastated by capitalist disinvestment.

Projects like Toledo's FGC and others involving brownfield conversions to renewable energy and urban agriculture, and redevelopment in general, align with the mission of Africana studies to contribute to the quality of the black experience and to study and report on that experience and its ongoing transformation. The quality of local citizens' lives is improved, and Africana scholarship become both more relevant in the community and more highly regarded on campus and in the academy.

Just as CDCs seek to foster greater social capital in black communities, where it is often in scarce supply, Africana studies can expand social capital between CDCs. In other words, it is ironic that CDCs tell residents that they have to work together to improve their environment, economy, and civic society, and yet they more often than not fail to work well with each other in the same city. Competition often trumps cooperation to the detriment of distressed community residents. Intra-city CDCs frequently are in alliance, but these alliances tend not to be very vibrant and productive.

There is something to be learned from the private sector in the way it simultaneously collaborates and competes. For example, a trade association supports an industry in its fight for lower taxes, less regulation, and greater subsidies. CDCs should also consider the corporate style of interlocking, in which directors and executives of one corporation serve on the boards of other corporations. The interlocking of CDCs, particularly those that are contiguous, would attract funding from sources such as the EPA and the DOE for brownfield remediation. If interlocked CDCs worked in contiguous service areas, the positive spillover from brownfield redevelopment could have a ripple effect.

Features of both urban agriculture and brightfields can be found in community solar gardens. Citizens may not have the time, skills, or land to grow their own gardens, but they can still consume produce grown in their neighborhood through an urban agricultural entity such as the FGC. Similarly, community solar gardens allow citizens to consume credit for solar energy without the up-front cost and the need for gardening skills or a roof on which to install solar panels. Community solar gardens are similar to brightfields in that solar energy can be generated on contaminated land without the most expensive forms of remediation.

Brownfields in Cities with Access to Africana Studies Programs

As derived from eBlackStudies.org, there are 262 strictly defined Africana studies programs nationwide, which are located in 207 cities that are home to a total of 4,146 brownfields. I used two well-known databases to determine the number of brownfields: the EPA's Superfund website (http://www.epa.gov/superfund/sites/index.htm) and the EPA's Cleanups in My Community website (http://www2.epa.gov/cleanups/cleanups-my-community). These databases listed brownfield sites for 184, or 89 percent, of the 207 Africana studies cities. For the 23 Africana studies cities not in the databases, I used various techniques to ascertain the brownfields in each one. One involved contacting the state or municipal environmental protection office, which yielded the number of brownfield sites for the 23 cities.

Figure 3.3 is a map showing the ten EPA regions. Figure 3.4 depicts the aggregate number of Africana studies programs for each region, and Figure 3.5 depicts the aggregate number of brownfields for each.

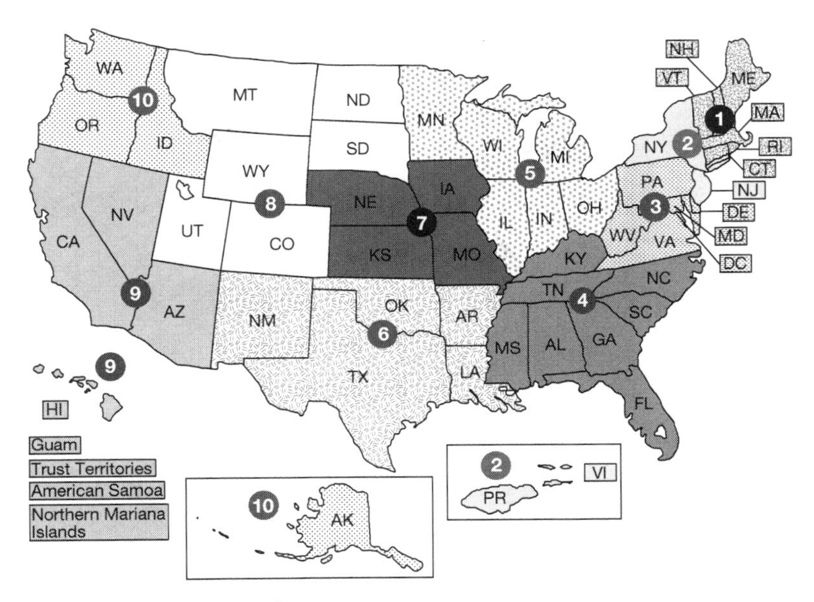

Figure 3.3 EPA regional map

Figure 3.4 Africana studies programs by EPA regions

Figure 3.5 Brownfields in EPA regions

We can see that region 4 has the greatest number, followed by region 5. Region 4 is known for having some of the weakest environmental laws and enforcement in the country, whereas region 5 is the nation's industrial heartland, also known as the rustbelt. Many of the brownfields in the Midwest are former sites of industrial and commercial facilities that moved to the suburbs, the U.S. South, or the Global South. Figures 3.6 and 3.7 respectively show the number of TRI facilities and Superfund sites in each EPA region. For each region, Tables 3.2 through 3.11 show the Africana Studies cities and their respective number of LULUs, which encompass brownfields, TRI facilities, and Superfund sites. (The last two columns in the tables, TRI facilities and Superfund sites, are discussed in the next section.)

In light of Africana studies' scholarly and social mission on the one hand and the health, environmental, economic, and civic problems posed by brownfields on the other, Africana studies programs can make substantial contributions to the black community and to the academy through rigorous research and community engagement on brownfield remediation and redevelopment modalities. If professors and students were to collect uniform data on the 4,164 brownfields

Figure 3.6 TRI reporting facilities in EPA regions

Figure 3.7 Superfund NPL sites in EPA regions

TABLE 3.2 AFRICANA STUDIES CITIES, PROGRAMS, AND LULUS: EPA REGION 1

State/city	Africana studies programs	Brownfields	TRI reporting facilities	Superfund NPL sites
Connecticut				
Fairfield	1	3	17	1
Mansfield	1	1	3	0
Middleton	1	24	14	0
New Haven	1	15	32	0
New London	1	5	11	1
West Hartford	1	4	24	0
State total	*6*	*52*	*101*	*2*
Maine				
Brunswick	1	8	4	1
Lewiston	1	23	6	0
Waterville	1	2	2	0
State total	*3*	*33*	*12*	*1*
Massachusetts				
Amherst	2	0	5	0
Boston	5	77	39	1
Cambridge	1	0	36	1
Chestnut Hill	1	0	25	1
(North) Dartmouth	1	4	13	4
Medford	1	6	47	3
Norton	1	0	0	0
Salem	1	0	0	0
South Hadley	1	1	23	0
Waltham	1	0	0	0
Wellesley	1	1	9	2
Williamstown	1	0	0	0
Worcester	1	20	28	12
State total	*18*	*109*	*225*	*12*
New Hampshire				
Hanover	1	1	0	0
State total	*1*	*1*	*0*	*0*
Rhode Island				
Providence	2	0	0	0
State total	*2*	*0*	*0*	*0*
Vermont	*0*	*0*	*0*	*0*
Region total	30	195	338	15

TABLE 3.3 AFRICANA STUDIES CITIES, PROGRAMS, AND LULUS: EPA REGION 2

State/city	Africana studies programs	Brownfields	TRI reporting facilities	Superfund NPL sites
New Jersey				
Ewing Township	1	68	20	0
Glassboro	1	0	0	0
Madison	2	8	19	1
Princeton	1	1	6	5
South Orange	1	1	58	7
Wayne	1	3	35	6
State total	*7*	*81*	*138*	*19*
New York				
Albany	1	31	12	1
Aurora	1	2	0	1
Binghampton	1	0	0	0
Brockport	1	2	0	1
Bronx	2	2	15	2
Buffalo	2	135	42	1
Clinton	1	1	13	0
Cortland	1	4	7	1
Garden City	1	0	0	0
Geneva	2	1	5	0
Hamilton	1	1	0	1
Hempstead	1	0	0	0
Ithaca	1	2	2	0
Jamaica	1	0	0	0
Lewiston (Niagara University)	1	0	0	0
New Paltz	1	2	0	3
New York	10	10	257	46
Oneonta	2	2	1	0
Plattsburg	1	4	2	1
Potsdam	1	1	3	0
Poughkeepsie	1	0	0	0
Purchase	2	0	1	0
Schenectady	1	0	0	0
Stony Brook	1	0	6	2
Syracuse	1	19	27	1
State total	*39*	*219*	*393*	*61*
Puerto Rico	*0*	*0*	*0*	*0*
U.S. Virgin Islands	*0*	*0*	*0*	*0*
Region total	46	300	531	80

TABLE 3.4 AFRICANA STUDIES CITIES, PROGRAMS, AND LULUS: EPA REGION 3

State/city	Africana studies programs	Brownfields	TRI reporting facilities	Superfund NPL sites
Delaware				
Newark	1	0	0	0
State total	*1*	*0*	*0*	*0*
District of Columbia				
Washington	2	55	123	12
State total	*2*	*55*	*123*	*12*
Maryland				
Baltimore	2	126	45	4
College Park	2	2	12	1
State total	*4*	*128*	*57*	*5*
Pennsylvania				
Bryn Mawr	1	0	0	0
Easton	1	0	0	0
Gettysburg	1	4	4	3
Haverford	2	0	0	0
Indiana	1	0	0	0
Lancaster	1	22	35	0
Lincoln University	1	0	0	0
Philadelphia	2	0	0	0
Pittsburgh	2	23	39	0
Randor Township	1	0	0	0
Shippensburg	1	0	0	0
State College	1	0	0	0
Swarthmore	1	0	0	0
State total	*16*	*49*	*78*	*3*
Virginia				
Blacksburg	1	3	5	0
Charlottesville	1	4	1	0
Fairfax	1	1	7	0
Norfolk	1	8	15	5
Richmond	1	83	18	0
Williamsburg	1	4	6	2
State total	*6*	*103*	*52*	*7*
Region total	**29**	**335**	**310**	**27**

TABLE 3.5 AFRICANA STUDIES CITIES, PROGRAMS, AND LULUS: EPA REGION 4

State/city	Africana studies programs	Brownfields	TRI reporting facilities	Superfund NPL sites
Alabama				
Birmingham	1	50	60	1
Huntsville	1	0	0	0
Montgomery	1	0	0	0
Normal	1	0	0	0
Selma	1	0	0	0
Tuscaloosa	3	0	0	0
Tuskegee Institute	1	0	0	0
State total	*9*	*50*	*60*	*1*
Florida				
Boca Raton	1	2	3	0
Coral Gables	1	0	0	0
Fort Lauderdale	1	0	0	0
Gainesville	1	20	7	1
Miami	1	402	11	7
Orlando	1	26	33	2
St. Petersburg	1	20	9	0
Tallahassee	2	18	5	0
Tampa	1	158	94	15
State total	*10*	*646*	*162*	*25*
Georgia				
Athens	1	0	6	1
Atlanta	5	85	49	0
Decatur	1	8	12	0
Milledgeville	1	3	1	0
Savannah	1	25	27	0
State total	*9*	*121*	*95*	*1*
Kentucky				
Frankfort	1	12	2	0
Lexington	1	14	7	0
Louisville	1	81	60	0
State total	*3*	*107*	*69*	*0*
Mississippi				
Oxford	1	0	0	0
State total	*1*	*0*	*0*	*0*
North Carolina				
Asheville	1	30	17	2
Charlotte	1	57	74	4

(continued)

TABLE 3.5 *(continued)*

State/city	Africana studies programs	Brownfields	TRI reporting facilities	Superfund NPL sites
Durham	1	63	13	2
Glen Raven	1	0	0	0
Greensboro	1	27	28	0
Wilminton	1	19	11	1
State total	*6*	*196*	*143*	*9*
Tennessee				
Knoxville	1	28	76	0
Memphis	1	28	79	4
Murfreesboro	1	0	0	0
Nashville	3	0	0	0
State total	*6*	*56*	*155*	*4*
Region total	**44**	**1,176**	**684**	**40**

TABLE 3.6 AFRICANA STUDIES CITIES, PROGRAMS, AND LULUS: EPA REGION 5

State/city	Africana studies programs	Brownfields	TRI reporting facilities	Superfund NPL sites
Illinois				
Carbondale	1	0	0	0
Champaign	1	3	11	0
Charleston	1	0	0	0
Chicago	4	66	357	4
DeKalb	1	1	6	0
Evanston	1	16	24	0
Macomb	1	1	0	0
Peoria	1	4	10	0
Springfield	1	0	0	0
State total	*12*	*91*	*408*	*4*
Indiana				
Bloomington	2	0	15	2
Greencastle	1	0	0	0
Muncie	1	0	0	0
Notre Dame	2	0	0	0
Terre Haute	1	14	0	0
Valparaiso	1	2	8	0
West Lafayette	1	0	0	0
State total	*9*	*16*	*23*	*2*

(continued)

TABLE 3.6 *(continued)*

State/city	Africana studies programs	Brownfields	TRI reporting facilities	Superfund NPL sites
Michigan				
Allendale Charter Township	1	0	0	0
Ann Arbor	2	14	8	0
Dearborn	1	30	37	2
Detroit	1	190	60	1
Flint	1	43	6	0
Lansing	3	0	0	0
Rochester	1	0	0	0
Ypsilanti	1	49	9	0
State total	*11*	*326*	*120*	*3*
Ohio				
Athens	1	2	0	0
Bowling Green	2	0	9	0
Cincinnati	1	30	57	1
Cleveland	1	40	113	1
Columbus	1	2	55	2
Gambier	1	0	2	0
Granville	1	1	11	0
Kent	1	0	0	0
Oberlin	1	0	1	1
Oxford	1	0	1	0
Toledo	1	143	47	0
Wilberforce	1	0	5	0
State total	*13*	*218*	*301*	*5*
Wisconsin				
Madison	3	19	23	1
Milwaukee	1	132	160	6
Oshkosh	1	34	17	0
White Water	1	2	6	0
State total	*6*	*187*	*206*	*7*
Region total	**51**	**838**	**1,058**	**21**

in cities where both programs and brownfields exist, they could create an invaluable database that the field would control. Local studies of LULUs in all cities with Africana studies programs, including national surveys and ethnographic work, would result in a richer body of knowledge that better reflects the situation on the ground, local

TABLE 3.7 AFRICANA STUDIES CITIES, PROGRAMS, AND LULUS: EPA REGION 6

State/city	Africana studies programs	Brownfields	TRI reporting facilities	Superfund NPL sites
Arkansas				
Fayetteville	1	0	6	0
State total	*1*	*0*	*6*	*0*
Louisiana				
Baton Rouge	1	0	0	0
New Orleans	1	34	14	1
State total	*2*	*34*	*14*	*1*
New Mexico				
Albuquerque	1	0	0	0
State total	*1*	*0*	*0*	*0*
Oklahoma				
Norman	1	2	5	0
Tulsa	1	5	107	2
State total	*2*	*7*	*112*	*2*
Texas				
Austin	1	48	27	0
Houston	1	99	363	15
University Park	1	0	0	0
Waco	2	0	0	0
State total	*5*	*147*	*390*	*15*
Region total	**11**	**188**	**522**	**15**

TABLE 3.8 AFRICANA STUDIES CITIES, PROGRAMS, AND LULUS: EPA REGION 7

State/city	Africana studies programs	Brownfields	TRI reporting facilities	Superfund NPL sites
Iowa				
Ames	1	4	7	0
Cedar Rapids	1	9	20	1
Grinnell	1	1	2	0
Iowa City	1	2	7	0
State total	*4*	*16*	*36*	*1*
Kansas				
Lawrence	1	1	8	0

(continued)

TABLE 3.8 *(continued)*

State/city	Africana studies programs	Brownfields	TRI reporting facilities	Superfund NPL sites
Wichita	1	154	48	4
State total	*2*	*155*	*56*	*4*
Missouri				
Columbia	1	5	6	0
Kansas City	1	0	0	0
Kirksville	1	1	0	0
Springfield	1	147	24	2
St. Louis	1	224	69	3
State total	*5*	*377*	*99*	*5*
Nebraska				
Omaha	1	19	26	1
State total	*1*	*19*	*26*	*1*
Region total	**12**	**567**	**217**	**11**

TABLE 3.9 AFRICANA STUDIES CITIES, PROGRAMS, AND LULUS: EPA REGION 8

State/city	Africana studies programs	Brownfields	TRI reporting facilities	Superfund NPL sites
Colorado				
Boulder	1	21	15	0
Colorado Springs	1	40	16	0
Denver	1	110	76	9
Fort Collins	1	14	7	0
Greeley	1	4	10	0
State total	*5*	*189*	*124*	*9*
Montana				
Missoula	1	7	4	1
State total	*1*	*7*	*4*	*1*
North Dakota	*0*	*0*	*0*	*0*
South Dakota	*0*	*0*	*0*	*0*
Utah	*0*	*0*	*0*	*0*
Wyoming	*0*	*0*	*0*	*0*
Region total	**6**	**196**	**128**	**10**

TABLE 3.10 AFRICANA STUDIES, PROGRAMS, CITIES, AND LULUS: EPA REGION 9

State/city	Africana studies programs	Brownfields	TRI reporting facilities	Superfund NPL sites
Arizona				
Tempe	1	1	61	2
Tucson	1	82	20	1
State total	*2*	*83*	*81*	*3*
California				
Berkeley	1	2	16	4
Carson	1	6	126	2
Chico	1	0	0	0
Claremont	5	4	20	0
Davis	1	2	2	2
Fullerton	1	0	0	0
Irvine	1	3	42	1
La Jolla	1	7	14	0
Long Beach	1	5	101	3
Los Angles	3	66	515	24
Sacramento	1	43	24	5
San Francisco	1	10	10	3
San Jose	1	21	67	24
San Luis Obispo	1	26	5	0
Santa Barbara	1	0	0	0
Stanford	1	0	0	0
State total	*22*	*195*	*942*	*68*
Hawaii				
Honolulu	1	0	0	0
State total	*1*	*0*	*0*	*0*
Nevada				
Las Vegas	1	0	0	0
State total	*1*	*0*	*0*	*0*
Pacific Islands	*0*	*0*	*0*	*0*
Region total	27	278	1,023	71

citizens' perspectives on associated problems, and more concrete and more insightful knowledge for community-based organizations and public policy decision makers. Moreover, these programs would be in a better position to win grants from agencies such as the EPA for funding brownfields research.

TABLE 3.11 AFRICANA STUDIES CITIES, PROGRAMS, AND LULUS: EPA REGION 10

State/city	Africana studies programs	Brownfields	TRI reporting facilities	Superfund NPL sites
Alaska	*0*	*0*	*0*	*0*
Idaho	*0*	*0*	*0*	*0*
Oregon				
Corvallis	1	4	4	1
Eugene	1	7	27	0
Portland	1	66	119	13
State total	*3*	*77*	*150*	*14*
Washington				
Cheney	1	10	1	0
Pullman	1	20	2	0
Seattle	1	47	42	8
State total	*3*	*77*	*45*	*8*
Region total	**6**	**154**	**195**	**22**

Toxics Release Inventory Facilities

Working to limit the harm caused by 4,164 brownfields to health, environment, and civil society in black communities where Africana studies programs are located, and to redevelop these sites, is enough to keep many Africana scholar-activists busy. But there is another major category of LULUs in the black community, TRI facilities (see Figure 3.8), that these scholar-activists can take a lead in addressing. As noted previously, the 1984 Union Carbide explosion that killed some twenty thousand Indians in Bhopal and severely injured some six hundred thousand others, along with smaller events in the United States, focused the American public's mind on the need to know what quantities of specific hazardous chemicals could cause a public health disaster. In 1986, Congress passed the Emergency Planning and Community Right-to-Know Act, which authorized the EPA to establish a reporting regime of the over 650 highly toxic materials that are manufactured, processed, and stored at thousands of facilities around the United States.

The EPA reported that in 2006 34 million metric tons of chemical substances were produced and imported daily. By 2030, the volume of

Figure 3.8 Toxics release inventory facility (TRI) in a low-income white community in Toledo, Ohio

chemical substances produced and imported is expected to have doubled (Rom 2011). In 2010, according to TRI data, 3.9 billion pounds of toxic chemicals were released into the environment. As Zachary Smith notes, "Cancer, nervous system damage, kidney, liver, chromosomal and lung damage, as well as genetic mutations are some of the impacts that may result from human exposure to toxic pollutants" (2012: 225). Without government regulation, enforcement, and monitoring, carcinogenic chemicals such as methanol, xylene, and toluene from the petrochemical industry (Hendryx and Fedorko 2011) and sulfur dioxide (SO_2) and nitrogen oxides (NO_X) from power-generating facilities, particularly coal facilities, would profoundly and irreparably damage human health and ecological systems.

On the government side stretching back to the 1970s, a complex set of monitoring and control regimes has been established to curb toxic releases that harm human and environmental health. Notes Richard Hula:

> More than two dozen pieces of federal legislation regulate toxic materials in the United States. Of these, three define the broad regulatory framework that has controlled toxic substances for the past twenty years. They are the Toxic Substance Control Act of 1976 (TSCA), the Resource Conservation and Recovery Act of 1976 (RCRA), and the Comprehensive Response, Compensation and Liability Act of 1980 (CERCLA). Each is targeted to a different period in the life cycle of hazardous material. TSCA attempts to set out rules for the review and analysis of new chemicals. The goal is to identify dangerous chemicals as they are developed so that appropriate control strategies can be devised before such chemicals cause human harm. RCRA charges the Environmental Protection Agency (EPA) to develop standards for current waste management. CERCLA is targeted to cleaning the nation's worst toxic waste sites. Together, these laws create a complex network of regulations that attempt to control toxic materials from inception to disposal. (2009: 184)

The 1970 Clean Air Act established National Ambient Air Quality Standards (NAAQS) for toxic pollutants deemed deleterious or deadly to public health and the environment. Among these are carbon monoxide (CO), nitrogen dioxide (NO_2), SO_2, particulate matter, ozone gas (O_3), and lead (Pb). CO, sulphides (S_X), nitrites (N_X), particulate matter, and ground-level ozone are the most common forms of air pollution. Many of the chemical substances that contribute to pollution were "grandfathered" into TSCA and therefore have been able to avoid toxicity testing.

We should think about the political dynamics of grandfathering for just a moment. The chemical industry fought the establishment of TSCA in order to keep new chemicals from toxicological analyses. Although it lost that battle, it won the next one by keeping the 62,000 or so chemicals already in use from testing. The industry also won another battle in the requirement that TSCA operate "under the default assumption that chemicals remain on the market unless or until EPA generates sufficient evidence to prove harm" (Rom 2011: 356). The chemical industry, with the aid of its chief trade association, the American Chemistry Council—as well as industry lobbyists such as the U.S. Chamber of Commerce and the National Association of Manufacturers—remains engaged in the political process to prevent and/or weaken regulation and enforcement. The onus is on the public to be even more politically engaged to protect public health and the environment above and beyond levels that the chemical industry may deem sufficient.

The TRI reports on the acquisition, storage, processing, manufacturing, release, and transfer of criteria chemicals at threshold volumes. As of 2012, there were 593 individual chemicals and 682 chemical categories on the TRI that require reporting. The TRI is far from being as muscular as the public may want, but it is what we have won in ongoing struggles with the chemical industry. Despite its limitations, it does provide a wealth of information for the public to use in efforts to curb toxic releases.

As mentioned previously, the blacker the census tract, the greater the likelihood that it hosts a brownfield. Likewise, according to John Tiefenbacher and Ronald Hagelman III, "counties with higher percentages of blacks and other minority groups are more likely to have higher numbers of acute and chronic toxic releases" (1999: 516). Thus, blacks must be even more politically engaged around the subject of harmful substances, since their communities are more likely to experience airborne toxic releases and hazardous material spillage. To test for residential proximity to industrial sources of air pollution in relation to race, poverty, and age, Susan Perlin, David Wong, and Ken Sexton (2001) closely examined three communities: Kanawha Valley in West Virginia, the Baton Rouge–New Orleans corridor, and metropolitan Baltimore. In all three, they found, African Americans lived closer to the nearest industrial emissions sources than whites, and they were more likely to live within two miles of multiple industrial emission sources. This study corroborates a large body of research showing that, nationally speaking and not just in selected communities, blacks are much more likely to live within two thousand meters of a TRI or Superfund site (Currie 2011). Jean Brender, Julian Maantay, and Jayajit Chakroborty conducted a comprehensive literature review "to examine the relation between residential proximity to a wide range of environmental hazards and cardiovascular and respiratory disease, PCB toxicity, end-stage renal disease, diabetes, and adult cancers such as leukemia and non-Hodgkin's lymphoma" (2011: S40). Interestingly, the investigators noted that "several studies have found that living near hazardous wastes sites, industrial sites, cropland with pesticide applications, highly trafficked roads, nuclear power plants, and gas stations or repair shops is related to an increased risk of adverse health outcomes" (S37).

Moreover, significant research shows an unmistakable relationship between school performance scores and proximity to toxic pollution sources, particularly that between developmental neurotoxins and school grounds. Cristina Lucier, Anna Rosofsky, and Bruce London uncovered a significant relationship between learning outcomes and proximity to "TRI releases of developmental toxins, suspected developmental toxins, suspected neurotoxins, lead, mercury, manganese, carbon disulfide, and toluene . . . located near 20 or more schools" (2011: 424).

Cristina Legot, Bruce London, and John Shandra, searched for similar relationships in a national proximity study that examined nearly seven hundred of the nation's highest-volume polluters of neurotoxins that pose the greatest health risks to children's health and learning abilities. They found "thousands of schools and hundreds of thousands of children at risk. . . . A substantial proportion of the high volume polluters studied are surrounded by disproportionate minority, poverty, and low-income populations" (2010). It is nearly impossible to overstate the vulnerability of children overexposed to pollution, who are, again, disproportionately black. Children exposed to severe air pollution can develop neuro-inflammation and structural alterations in the brain resulting in cognitive deficits (Calderon-Garcidueñas et al. 2011).

During the Bush administration, between 2006 and 2008, the EPA introduced "burden reduction" rule changes to the TRI. Threshold reporting requirements for chemical releases were raised from five hundred to two thousand pounds and created an initial threshold of five hundred pounds for persistent bioaccumulative toxins (PBTs). OMB Watch (2006) reported that the EPA received over 122,400 comments from individuals and organizations regarding the rule changes, 99.97 percent of which were against them. The "burden reduction" filings did not affect all TRI facilities equally. The major toxin-releasing facilities allowed to spew more toxins into the environment were disproportionately located in black and poor communities (Miranda, Keating, and Edwards 2008).

Companies with chemical releases in excess of established thresholds now had to file a Form R, whereas those with only criteria chemicals at their facilities had to file a Form A. According to Marie Miranda and colleagues' findings:

> The TRI Burden Reduction Rule does in fact have a disproportionate impact on minority and low-income communities because facilities eligible to file Form A under the new rule are more likely to be located in neighborhoods where the proportion of minority and low income residents is significantly higher than neighborhoods with facilities still required to submit Form R. Thus, low income communities and those of color appear to be losing disproportionately more information under the TRI Burden Reduction Rule. (2008: 5412)

But as is well known, elections have consequences. In 2009, the Obama administration eliminated the "burden reduction" rule, returning to the previous reporting requirements.

It should be pointed out that black children are also disproportionately adversely affected by vehicle-related pollutants. Because communities of color and high-poverty neighborhoods have over twice the level of traffic density as white and more affluent neighborhoods, they have higher localized concentrations of "vehicle-related pollutants associated with chronic illness, lung impairment, and increased morbidity and mortality" (Houston et al. 2004). Many studies show significantly higher levels of childhood cancer, leukemia, low birth weight, and asthma in neighborhoods in high-traffic areas. Conversely, affluent whites and affluent individuals in general who reside in upscale condominiums in places such as downtown Chicago, where there is a substantial amount of vehicle-related pollution, do not suffer the same poor health. They have better nutrition, excellent "preventive and treatment" health care, less stress, and more leisure time, all of which mitigate threats to wellness.

Chapter 5 discusses air pollutants and climate change in Africa, another issue that should be taken up by Africana studies. A major source of African air pollution is biomass burning, which is the principal source of energy for cooking and heating for Africans, most of whom are not connected to an electric grid. Burning often takes place indoors—hence the disease known as "hut lung," a lower respiratory tract infection caused by black soot. In addition to undermining the health of millions—primarily women and girls—and causing the premature annual death of over two million around the world, biomass burning contributes to greenhouse gases through the cutting of trees that absorb carbon, which is released by burning.

Toxics Release Inventory Facilities in Africana Studies Cities

The 207 cities with Africana studies programs, as provided by *eBlack Studies*, are home to 5,006 TRI facilities and 335 Superfund sites. These numbers are more precise than the number of brownfield sites because the tallies of brownfield sites are compiled by each jurisdiction, often using different criteria and varying levels of consistency and rigor as well as varying levels of commitment to accurate documenting and reporting. That said, it should be noted that actual numbers of TRI facilities and Superfund sites are not mathematically precise, either. Documenting TRI facilities depends on self-reports, which are undoubtedly tempting to skew for some companies with no qualms about chicanery. Superfund site documentation can involve a degree of politicking as well. Some jurisdictions might attempt to politically manipulate the EPA's definition of a highly contaminated site as a Superfund site if they think it might result in greater federal funding for remediation. Similarly, some jurisdictions might apply political pressure to prevent a site from being designated a Superfund site if they think it would harm potential investment and if they are convinced that very little, if any, federal funds would be invested to remediate the property. The point here is not to focus on exact numbers but rather to focus on the aggregate. In other words, the ratio of environmental contaminants to GDP has been declining, and one cannot deny that TRI reporting and control in general have contributed to a cleaner environment per unit of economic activity than what might have been without them.

It is difficult to overstate the degree to which industrial polluters such as the chemical and petrochemical industries struggle to reduce transparency, regulation, and enforcement. On the other side, the public wants more transparency, stricter regulation, and stiffer penalties for violations, and to get these it must become better informed, more engaged, and relentless in pressing the government to counteract the overwhelming government-capturing approach used by industrial polluters.

Tables 3.2–3.11 show that region 5, the nation's industrial heartland, has the largest number of TRI facilities in Africana studies cities at 1,058, followed by region 9 at 1,023, with California having 97.5 percent of the region 9 total. As for the region with the greatest number of Superfund sites in Africana studies cities, region 2 stands out with nearly twice as many as second-place region 6. The numbers of TRI facilities and Superfund sites in each Africana studies city for each of the ten EPA regions are also presented.

This chapter demonstrates not only that Africana and environmental studies should be working together but also that the mission of neither can be achieved without help from the other. Chapter 2 focuses on perspectives, while this chapter makes environmental concerns much more concrete. The quality of the environment is often a matter of life or death in the black community, but it is much more often a matter of increased morbidity and stunted cognitive ability leading to reduced educational performance and attainment and thus diminished economic opportunity. Chapter 4 discusses some career opportunities in remediation of environmental degradation and its attendant quality-of-life issues in black communities and in broader society, as well as opportunities in developing both a sustainable economy and sustainable communities.

In the early 2000s, well-meaning institutions and intellectuals held the unrealistic belief that there would soon be an abundance of green jobs that could soak up unemployment in black and other communities where high unemployment was stubbornly persistent. That belief has all but disappeared. Analysts are now much more sober in their discussions about green jobs in general and the black community in particular. Still, it must be said that a lack of abundant green jobs is not the same thing as no green jobs. Chapter 4 discusses some existing and potential green jobs and the training programs and curricula in joint Africana–environmental programs that can prepare students for them.

VIGNETTE 3.1. E. MICHELLE MICKENS

Community-Based Organizations Need Your Help on the Environment!

It really wasn't that long ago that I was disconnected from the way environmental injustices affected a community. Like many of the residents I serve, the concerns I had pertained to social injustices and developing affordable housing. In 2005, I had an epiphany when I began struggling to find adequate land to redevelop. I stumbled across brownfield redevelopment and was astonished by the number of abandoned factories, warehouses, and industrial sites in the community I serve. In 2009, the City of Toledo's Department of Environmental Services identified twenty-one brownfields within a three-mile radius between I-75 North, Bancroft Street, Detroit Avenue, Collingwood Boulevard, Hamilton Street, and Nebraska Avenue, which is the heart of the black community and represents most of the central city. The magnitude of the continuous exposure to high levels of airborne contaminants that compromise children and seniors' breathing capacity is overwhelming. Soil becomes polluted by the leeching of residues from chemicals simply left or improperly removed from abandoned industrial sites. The residual toxins create an environmental threat and adversely impact quality of life for residents living in the community.

Because of the many vacant lots in this community, finding appropriate ways to reuse them has been challenging. Most of the lots were unbuildable, meaning that houses or businesses could not be built on them because either the buildings were prohibited by the shape and narrowness of the lot or the lot was too expensive to clean up. Over the past six years, while serving as director of the Toledo Community Development Corporation, transforming vacant lots into community gardens and urban agriculture sites became very popular in neighboring cities, so I decided to investigate how we could incorporate this type of land reuse into the community we serve. Our agency does brownfield redevelopment, transforming abandoned, vacant, and blighted areas into sites of urban agriculture.

Food deserts are common in most urban communities where there is no decent grocery store that offers fresh, healthy, and affordable produce. Our community is suffering numerous diseases because of lack of proper nutrition. Yet many of our young female heads of household are convinced there is no need to improve the quality of the food they and their children consume. Generations of poor eating habits persist throughout our neighborhoods, and community members are silent because they don't know or don't care about the impact their choices are having on their lives.

So we, a small neighborhood-based organization, joined with other stakeholders and community advocates who share a common vision. Our goal is to take transformative steps to influence the decisions our residents make concerning their livelihood with the intent of contributing to the improvement of their quality of life. We partnered with the Africana studies program at the University of Toledo under the leadership of Rubin Patterson, the Martin Luther King Academy for Boys, Robinson Elementary School, the City of Toledo, churches, and other nonprofit organizations to expose injustices and to restore opportunity for this struggling, predominantly black community. Although some days working at Toledo CDC are more stressful and discouraging than others—so much effort is required for small, almost imperceptible neighborhood improvements—you appreciate your work when reflecting on the fact that, without us, the community would have more urban blight, more civic disengagement, less employment, and less neighborhood pride. We have been responsible for bringing together skilled talent from across metro Toledo as well as attracting federal, state, and local dollars into the community. Again, the absence of our CDC would have resulted in much more decay.

Urban gardens, brownfield redevelopment, and retrofitting/rehabilitating existing structures will continue to be our stock in trade as Toledo CDC moves forward. These strategies will result in a healthier environment and a healthier citizenry that is imbued with economic vitality. Our green approach will result generally in green life-skilling and technically in green-skilling local citizens so that they can be employed in improving the community. Africana studies programs at CDCs around the country can always play a monumental role, just like the one the University of Toledo played, by assisting with everything from conducting social science research to providing students for internships to executing program evaluation. Local citizens of economically distressed black communities need CDCs, Africana studies, and other entities working collaboratively to provide them with opportunities to pursue their dreams and to perform up to their potential.

VIGNETTE 3.2. BEATRICE MIRINGU

Making the Environment Your Life's Work

A solid academic background in biological sciences is a door to a dynamic career in the diverse field of environmental stewardship. With a passion for protecting the environment and a drive for championing sustainable use of natural resources, one can be rewarded with a career that is environmentally indispensable, socially crucial, intellectually challenging, and emotionally fulfilling.

I was born in Kenya and received my B.S. in botany from Kenyatta University. After graduating, I joined the National Museums of Kenya as an assistant research scientist and worked on several projects, including the Kenya Indigenous Forest Conservation Project and the Plant Conservation Unit for rare and endangered species through the East African Herbarium. In both, I participated in ecological surveys to document species diversity and population status and in gathering socioeconomic data on how local communities utilize natural resources. I embraced the belief that indigenous knowledge sustains the balance between resource extraction and ecological preservation.

By observing and documenting communities' interactions with their natural resources, I became inspired to enroll in the Miami University Institute of Environmental Services in Ohio, focusing my graduate work on local community participation in conservation and utilization of natural resources. In pursuing graduate studies, my eyes were opened much wider to the burden of industrial pollution on communities living in close proximity to manufacturing facilities. The economic benefits that such facilities bring to the community often lead many to acquiesce and accept the tons of toxic pollutants that adversely affect residents' health. While my initial interest was working with local communities on natural resources management, I felt the need to work with an organization focused on pollution prevention.

In 2002, I joined Ohio Citizen Action, a statewide nonprofit organization, to work on a Good Neighbor campaign in Northwest Ohio. Good Neighbor campaigns are organized on the philosophy that businesses and communities share the burden of protecting neighborhoods from pollutants. My duties included strategic campaign development, door-to-door community and grassroots organizing, and fund-raising. I found out that fund-raising for grassroots environmental initiatives requires being technically sound and communication savvy so as to narrate a compelling story quickly and persuasively. Other related strategies used to educate the public included working with news media and organizing community meetings. These are ideal skills that liberal arts fields such as Africana studies, sociology, and women's and gender studies provide.

During campaigns, I partnered with local environmental activists in educating Northwest Ohio communities on the disastrous corrosion problem at the Davis-Besse nuclear power plant, located on the shores of Lake Erie near Toledo. These activities led to opportunities to collaborate with the National Refinery Reform campaign and the Environmental Integrity project.

From 2005 to the present, I have been working with the City of Toledo, Division of Environmental Services, where my primary focus has been enforcement of and compliance with environmental protection codes. Specifically, my daily activities ensure that the City of Toledo, businesses, and citizens comply with the National Pollutant Discharge Elimination System (NPDES) and other federal, state, and municipal codes. I work with local communities through public education and outreach programs. One such program is the Toledo–Lucas County Rain Garden Initiative, which advocates the use of green infrastructure in surface runoff management.

I have also contributed my skills and expertise to international sustainable development projects. For example, I have worked with the Coalition for a Sustainable Africa (CSAfrica) on a tree-focused sustainable development project in West Africa. This project assists local farmers in growing Jatropha plants, which have high yields of biodiesel.

My latest and most exciting projects have involved working with Dr. Rubin Patterson on brownfield redevelopment research projects in local CDC territories and teaching an Introduction to Sustainability class in the University of Toledo Africana studies program. Through this class, students are made aware of environmental justice issues, corporate social responsibility, and the basis of the Clean Air Act and Clean Water Act, which constitute the backbone of environmental standards and accompanying enforcement and compliance requirements for pollution control. Throughout my career, I have found this field to be very dynamic and the ability to switch hats very rewarding. An excellent liberal arts education in Africana studies provides the intellectual fluidity and greater flexibility necessary to more easily adapt to changing circumstances.

For me, it all started with a solid foundation in the biological sciences as an undergraduate. I was initially inspired to study environmental sustainability by my fellow countrywoman, the Nobel laureate and wonderful Wangari Maathai. Her brilliance, magnetism, and determination to achieve environmental sustainability with social justice captured my imagination and never let go. Africana studies is one of the spaces where students learn and are inspired about the phenomenally important work by people of African heritage such as Dr. Maathai. When you combine the two, environmental science and Africana studies, you then have a powerful education.

4/ Green Jobs

Liberal Arts Fields Support the Growth of Green Jobs, Too: Everything Is Political!

In this age of neoliberalism and restricted pathways to upward social mobility, leaders of higher-education institutions are increasingly privileging science, technology, engineering, and mathematics (STEM) at the expense of liberal arts. Erick Fingerhut, former chancellor of the Ohio Board of Regents, argues that schools of higher education should focus on programs that create jobs and on course offerings that are clearly linked to labor demands. He is not only giving voice to the sensibilities of private-sector enterprises in Ohio but also reflecting the sentiment of students attending or soon to be attending one of Ohio's state universities and their parents. All state college/university chancellors subscribe to versions of Fingerhut's vocationalizing of college programs.

As Bakari Kitwana (2005) points out, the "get money" notion is as central to the hip-hop generation as social uplift was to the civil rights generation and as central to the millennials (those born after 1980), who are often regarded as the narcissistic "me generation." Although pioneers of Africana studies four decades ago did not see themselves as working to produce careerists, African American and other college students in Africana programs are not unlike all college students in the general sense that they want to complete their studies and launch status-laden and materially rewarding careers. A more realistic goal for Africana studies is that its graduates pursue rewarding and fulfilling careers with an impulse toward social justice—including engagement with environmental matters. Green Africana programs provide abundant opportunities for both.

Just like other social science and humanities programs, Africana programs have an important though perhaps not easily perceptible role in greening the economy and promoting a sustainable society. Nevertheless, they have been underemphasized in engagement of what is arguably the gravest challenge to all humanity—worldwide environmental degradation and its cure, sustainability. STEM and business are thought of primarily when it comes to sustainability, as they are associated with discovering new secrets of nature, pioneering new technologies, and establishing new business models for corporate profitability and jobs. However, Africana studies, social sciences, and the humanities have roles to play that are infinite, complex, and absolutely indispensable to the success of environmental sustainability, although many institutional leaders have yet to appreciate this important point.

The thread connecting the social sciences, humanities, STEM, and business can be illustrated by California's Global Warming Solutions Act of 2006, the goal of which was to bring greenhouse gas emissions back to 1990 levels by 2020 and to promote economic development and job growth. To reach these goals, new industries, companies, and technologies would have to be developed to achieve low-carbon emissions in both manufacturing and consumption in a state that in 2006 was the world's eighth-largest economy. The broader intent of the act would be the eventual creation of a green Silicon Valley.

Oil refineries and other major industrial polluters were steadfastly against the act and in 2010 backed Proposition 23 in an effort to suspend it until California's unemployment—over 12 percent in 2010—was reduced to no more than 5.5 percent for four consecutive quarters. Given that it had been decades since California had experienced such low employment numbers, this was undoubtedly a ruse. Texas-based oil companies Valero (the nation's biggest independent oil refiner) and

Tesoro were Prop 23's principal benefactors, and 97 percent of contributions toward its passage came from oil-related companies such as Koch Industries. Major corporations supporting the Global Warming Solutions Act included Pacific Gas and Electric and Nike.

Because Californians cared about climate change, yet cared even more about jobs in the face of high unemployment at that time, propolluters made the case that Prop 23 would protect jobs. In a sense, informed Californians wanted their economy to model that of Denmark, which had grown by 78 percent over the last three decades, even with a reduction in carbon dioxide emissions. In other words, they believed that economic growth and carbon emission reductions could occur simultaneously.

In spite of its heavyweight backing, Prop 23 was defeated, so the "technology push" in response to the "demand pull" from the Global Warming Solutions Act will no doubt continue to foster innovation in part through state funding for green research and development, venture capital investment, and the Silicon Valley culture of technology-based entrepreneurship. The act has also created many jobs for engineers and scientists and for professionals who work in clean business fields such as management and finance. After Prop 23 was defeated, even ExxonMobil, a leading supporter of climate change deniers, announced that it "would invest $600 million with a La Jolla biotech firm to create fuel from algae" (Roosevelt 2010) in response to the act, which would remain in effect. Clean tech firms are now promoting a green economy as a result of the act and its successful defense, which significantly depended on a large number of educated people outside of the STEM and business fields—namely, those with training in the social sciences and the humanities. Supporting organizations included the Ella Baker Center, the Clean Energy Hip-Hop Tour, the National Wildlife Federation, Communities United, the California Alliance, CALPIRG, California Students for Sustainability, the Sierra Club, Credo Action, and many others. Numerous engineers and scientists who are now plying their trades, and many clean-tech companies that are earning revenue, owe their good fortune in part to the humanists and social scientists who worked cleverly and effectively to save the Global Warming Solutions Act, many of whom are employed in clean technology and other greening corporations, not merely in nongovernmental and community-based organizations (NGOs, CBOs) or government agencies. The point is that social scientists and humanists, as well as scientists and engineers, have roles to play in promoting a sustainable economy.

Citizens with a liberal arts education can be instrumental in the transition from a brown to a green economy through their unique contributions to united labor-environmental campaigns such as the Blue-Green Alliance. Collaborations of this nature were witnessed in the Global Warming Solutions Act campaign. In the past, organized labor regularly sided with corporate polluters against environmentalists because big industrial unions helped to deliver higher wages and better working conditions in some of the dirtiest industries, including mining and fossil fuels and steel and chemicals. Nonunionized labor benefited as well. However, the calculation has changed significantly as labor and environmentalists unite in efforts to create more and better green jobs and a healthier environment. Unions historically have been situated on the spectrum between employers and environmentalists. Valero and other backers of Prop 23 were clearly appealing to workers' concerns about loss of jobs, which has kept labor from asking for too much relief from occupational health hazards and community contamination. The weapon employed is known as environmental job blackmail.

Another reason behind the old labor-environmentalist rift is class. Historically, organized labor's education levels ranged from less than high school to some college. Traditional environmentalists, on the other hand, tended to be highly educated, as discovered in a 1969 national survey of Sierra Club members that found that 74 percent of the members had at least a college education. A 1972 study revealed that nearly 60 percent of environmental volunteers held college or graduate degrees, which was about five times greater than in the general population at the time (Taylor 2002). The priorities of trade unionists and environmentalists were often perceived as at odds, keeping the two from uniting.

Fortunately, in 2009 the labor-environmentalist dynamic was radically better than it had been in 1979 or even 1999. The two groups are now talking a common language and are experiencing something of an identity consolidation. The so-called blue-green coalitions (blue for labor and green for environmentalists) are not just marriages of convenience but are becoming marriages for the long haul. With brown unionized jobs drying up, it is not surprising that labor is siding increasingly with environmentalists; both envision new employment prospects in renewable energy, reduced pollution, and energy efficiency.

The collective identity of labor leaders, environmentalists, and social justice advocates has been demonstrated over the past few years in campaigns for clean energy legislation, stiffer environmental regulation, and economic stimulus policies. Green dimensions of Africana studies and other liberal arts disciplines are indispensable in the education and mobilization of activists for policies that accelerate the move toward a green economy.

Shifting to a Low-Carbon Economy Creates Jobs

The United States is incrementally shifting to a low-carbon economy, becoming a little more green in emissions and toxicity levels per unit of production. This shift is primarily a result of four factors: (1) physical changes in the environment, (2) successful prodding of public policy and regulation, (3) technological innovation, and (4) new markets for greener products and changing consumer habits.

Green restructuring engenders new job opportunities while eliminating older ones. In other words, "new jobs are not necessarily created in the same sectors, regions and communities where old jobs have been lost" (Strietska-Ilina et al. 2011: 61). Thus, although transitioning from a brown to a green economy will result in new jobs, specific policies are needed to reskill and reposition workers for the switch. Consider the efforts to decrease coal as a share of the nation's power portfolio and replace it with renewable energy, along with government policies restricting coal power, technological innovations that bring renewable energy closer and closer to grid parity with baseload capacity, and long-term growth trends for renewables.

Workers in the coal power industry are understandably less optimistic about the switch from coal to renewables because of the loss of employment. The Beyond Coal campaign, led by the Sierra Club, has been successful in imposing a de facto moratorium on new coal power plants, but it has only recently begun to address the resulting job losses. Although far more renewable energy jobs will eventually be created than presently exist in all fossil fuel industries, this might be of little comfort to coal power facility employees who are losing jobs now. For this reason, smart policies are needed to provide the necessary escape velocity for the move to renewables. Consider that natural gas extracted by fracking may slow the growth in renewables because the energy from fracked gas is more immediately available, and like other fossil fuels, it has a great deal of political support in Washington—in fact, far more than renewables at this point.

Let's look briefly at the four factors driving the progression to green cited previously: physical changes to the environment, government policies and regulations, technological innovation, and markets for green goods. Changes to the physical environment are already baked in, as McKibben (2010) notes regarding the alteration of planet Earth to what he calls "Eaarth." In other words, the planet on which McKibben was born half a century ago has changed in some fundamental ways. Melting polar ice sheets and mountain caps, increasing water scarcity, reduced biodiversity, and weakening ecosystems are phenomena that have already occurred and will continue to spawn chaotic environmental changes irrespective of how green we become. We can, of course, decelerate the damage, but we cannot stop it, because the momentum already triggered will have to play itself out. The aggregate physical, chemical, and ecological outcomes of environmental damage will engender social, political, and economic changes. And we are already seeing countless examples of this, one being environmental refugees. These menacing changes constitute one impetus behind the ongoing greening efforts in many societies.

Unsustainable consumption in the developed world and jarring high-carbon growth in the developing world demand either a collective decision to consume less (don't hold your breath!) or pioneering new green technology that will meet people's needs without accelerating harm to the environment. To put a finer point on it, since Fatih Birol (2006), chief economist of the International Energy Agency, and others have noted that global energy demand will rise by 50 percent in less than two decades, the global climate and the world's ecosystems, which are already unstable and weakened by fossil fuel energy and emissions-intensive production, can easily be predicted to further atrophy. Not only are the 1.4 billion people currently unconnected to an electric grid seeking that connection, but the billions more being born will seek it as well. Half the world's population lives in urban areas, and it accounts for 60 to 80 percent of the world's energy consumption and 75 percent of global carbon emissions (United Nations Environment Progamme, n.d.).

Gus Speth, cofounder of the Natural Resources Defense Council and former head of the United Nations Development Programme, argues for decoupling job growth from economic growth. Because of the fragility of the environment, he says that we need a post-growth society that generates many well-paying jobs using renewable energy and, even more important, that continuously accelerates efficiency improvements. As Speth put it, "We're going to have to 'de-grow' a lot of things to live within the planetary boundaries" ("Gus Speth" 2011: 6).

Speth goes beyond what other forward-thinking people have proposed, which is to delink pollution from production. Scholarly research on this topic in major journals, such as the *Journal of Industrial Ecology*, has focused on such technological and policy feats, as have some business and industry efforts. Speth's notion of delinking jobs from production is far ahead of those of today's progressive sustainability thinkers. But then, Speth has been ahead of his time for decades on environmental issues, from making the case for a White House Council on Environmental Quality and serving as its inaugural chairman to cofounding the National Resources Defense Fund. Before long, we are likely to see journals, a community of scholars, and industrial opportunities revolving around the type of delinking he is proffering. In public green policy, the broadest approach would be that of the Global Green New Deal (GGND) called for by the United Nations Environment Programme (UNEP). The GGND would allocate at least 2 percent of global GDP to clean energy and other green economic activities. This represents $1.4 trillion per year over a ten-year period, which is considered necessary to achieve a worldwide green economy.

In the United States, we saw such unprecedented spending with the 2009 American Recovery and Reinvestment Act (ARRA), signed into law less than one month into the Obama administration. The law resulted in $787 billion worth of spending, more than $40 billion of it earmarked for environmental projects and green job training. President Obama has pursued green policies (which should be rapidly scaled up) that have more than tripled spending on clean energy research, development, and commercialization. Green public spending is much higher in Asia: China spends more than any other nation, and South Korea allocates more than any other nation in terms of percentage of overall public spending.

We can say at least two things about U.S. government spending on green economic activity. First, it will affect green jobs in the aggregate as well in their geographical and economic-sector distribution. Knowing how green public policy drives the creation of green jobs and the destruction of brown jobs geographically as well as by economic sector can improve the prospects of African Americans in gaining their share of green jobs. Second, government spending to create green jobs is subject to political pressure. Consider California's Global Warming Solutions Act. Powerful interest groups such as the fossil fuel industry that are against faster economic greening compete with other interest groups that seek to profit from it, such as renewable energy industries, and with environmental social justice groups that advocate for more green jobs by shaping public policy. African Americans' active involvement in the political process can help shape public policy for more green jobs, but once more have been created, they still have to have the right skills and credentials to effectively compete for them.

Public support of weatherization programs creates many jobs in low-income communities. According to the U.S. Department of Energy, for every $1 million invested in weatherization programs in low-income communities, fifty-two jobs are created (U.S. Department of Energy 2002). Surprisingly, it was George W. Bush who in 2007 signed the Energy Independence and Security Act, which included the Green Jobs Act authorizing $125 million to be spent on training nearly thirty thousand workers for jobs in emerging green sectors (Harper-Anderson 2012). Using studies conducted by the Economic Policy Institute, the Apollo Alliance, and other organizations, Elsie Harper-Anderson concludes that more jobs from green investments are occupied by workers without a bachelor's degree than with one, while workers in those jobs earn wages some 25 percent higher than those of workers in nongreen jobs requiring similar education. More green policies with an eye toward jobs and sustainability would result in many jobs benefiting low-income African American communities. Brownfield remediation, building retrofitting, green construction, and green public transportation would arguably rank among the initiatives most likely to affect employment statistics. For example, energy auditing is a well-paid, growing job opportunity for which African Americans can qualify rather quickly. Environmental auditors assess the compliance of businesses with environmental legislation and standards (Strietska-Ilina et al. 2011).

The Apollo Alliance and the Blue-Green Alliance are among the best-known green jobs organizations in the country. Both work with businesses large and small and with governments at every level to gain support and funding to accelerate green jobs creation and training (Deitche 2010). This makes them exactly the type of institution that local Africana studies programs should be partnering with. When local Africana studies programs also bring in CDCs and community-based organizations, an important broad-based pressure group/constituency is formed for green public policy and hence more green employment. Africana studies can unite with these organizations in joint and contract research that supports the case for greener public policies and thus greener jobs for blacks and cleaner environments for their communities.

In *Skills for Green Jobs* (2011) Olga Strietska-Ilina and coauthors describe the bottlenecks in green job creation that can be traced to ineffective public policy: many individuals have completed green training programs but have not gained green employment, and yet a significant number of green jobs go unfilled. Figure 4.1 shows how green jobs grow both quantitatively and qualitatively. Some grow quantitatively with no additional green skill required. These include forest ranger jobs stemming from government-led expansion of environmental reserves and public transportation jobs because of the transition from high-emission to low- or no-emission vehicles. At the other end of the spectrum is the growth in jobs in clean-tech product and process innovation that require highly technical skills such as green chemistry. Most jobs, however, are medium skilled. That said, new occupations usually require higher qualifications, and according

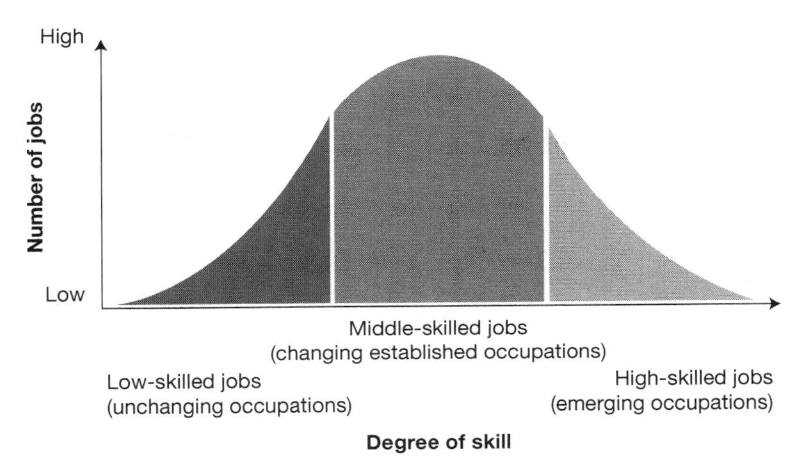

Figure 4.1 Relationship between education and number of jobs

to one study, "in 75 per cent of companies 'greening' generated a need for higher-skilled workers" (Strietska-Ilina et al. 2011: 23). Figure 4.1 depicts the distribution of green jobs by skill level.

For an illustration of how political outcomes affect public policy, which influences the pace of technology and industry greening, we can examine the Department of Energy's long-awaited funding of the Advanced Research Projects Agency–Energy (ARPA-E), which is the Department of Energy's version of the Department of Defense's iconic Defense Advanced Research Projects Agency (DARPA). DARPA was established a half-century ago to expand the frontiers of military technology for the Pentagon. Its research led to the commercialization and wide expansion of computer chips and the Internet into civilian applications, which has totally revolutionized how we live. ARPA-E was enacted in 2007, but the Bush administration, in the fossil fuel industry's interest, simply refused to have it funded. ARPA-E advocates believed that it would revolutionize innovation in renewables and other sustainability industries. It finally received its first $400 million appropriation from the Obama economic stimulus package for investments in transformational energy research projects or high-risk but potentially high-reward renewable energy technologies that the private sector cannot afford. Without a doubt, funding for ARPA-E needs to expand multifold in short order.

In his testimony to the U.S. House of Representatives in fiscal 2013 budget hearings, Dr. Arun Majumdar, director of ARPA-E, stated:

> Our grid is an aging infrastructure. Here is an example. The average age of a transformer in the United States is 42 years, 2 years beyond [its] projected lifespan. A typical 1 Megawatt (MW) transformer in our distribution substation weighs 8,000 lbs and is manufactured by foreign companies. An ARPA-E awardee, Cree based in North Carolina, is creating a quantum leap in electrical power technology. They are developing a 1 MW transistor made of silicon carbide the size of your fingernail. If they are successful, the 1 MW transformer could then shrink from 8,000 lbs to 100 lbs with greatly reduced cost and increased reliability. Because the U.S. is the world's leading manufacturer of silicon carbide, the Cree project could transform future electrical power technologies and create a large export market. (Advanced Research Projects Agency–Energy 2012).

The Obama administration, which champions ARPA-E, did not allow threatened polluting industries to prevent funding of the new agency. However, these industries did manage to severely limit it. The fiscal year 2013 funding request was a mere $350 million. Given the potentially game-changing innovations that ARPA-E-funded research, development, and commercialization are already producing, its annual budget should be in the tens of billions rather than the hundreds of millions. However, such a large-scale ramp-up in funding would require a large enough ramp-up in progressive political energy to elect a president and a Congress that prioritizes green public policy.

Venture capital (VC) is also investing in green research and development, hoping for blockbuster innovative products that will revolutionize markets and further enrich VC investors. A few of the major VC investors in green products and processes are Kleiner, Perkins, Caufield and Byers; Khosla Ventures; Vantagepoint Capital Partners; and Draper Fisher Jurvetson. Not surprisingly, these companies were major supporters of California's Global Warming Solutions Act and vigorous opponents of Prop 23. Over the past five years, green technology has vaulted to the inner circles of VC funding, along with information technology (IT) (e.g., cloud computing, new social media, software, wireless telecom) and biotechnology. Venture capital that is backing clean tech innovation has experienced the fastest growth in recent years. "Since 2002, the green sector has spawned a greater share of gazelles than the economy overall," note Karen Chapple and colleagues (2011: 13). "Gazelles" is business-speak for companies that start from a base of at least $1 million and experience revenue growth of at least 20 percent for four years. Venture capitalists are constantly on the hunt for gazelles, and such high-growth companies are increasingly found in the clean tech arena.

At an investment summit on climate risk in New York City in January 2010, over five hundred business leaders representing financial, corporate, and VC entities participated in discussions about investing in climate change solutions. Those participating have combined assets of some $20 trillion, or about one-third of the entire global economy. The conference made clear that many major businesses appreciate that pollution is triggering disruptions to weather patterns and other environmental phenomena, and they appreciate the profit potential from greening. This is why green job and sustainability advocates want to work with green-leaning corporations such as these, notwithstanding the often monumentally different motivations for going green.

According to Anthony Cortese in *Green Jobs for a New Economy*, "We need to redesign the human economy to emulate nature, operating on renewable energy and creating a circular production economy in which the concept of 'waste' is eliminated because all waste products are raw materials or nutrients for the industrial economy" (2009: 8). Africana studies graduates with a foundation in sustainability can go on to acquire higher-level policy-making skills so as to design policies to further green the broader economy and create employment in cleaner black communities.

Increasing green VC, ARPA-E, and other support for a green economy is among the reasons that, according to the U.S. Conference of Mayors (2008), the Bureau of Labor Statistics predicts green job growth of more than four million by the year 2038. As impressive as that sounds, it might be overly conservative. For instance, the U.S.

Energy Information Administration, which collects, analyzes, and disseminates energy information to policy makers, businesses, and the public, predicts that between 2010 and 2035 the renewables component of the nation's electric power portfolio will grow from 9.1 to 17 percent, while the coal component will shrink from 48.5 to 43.8 percent (Atlantic Council 2010). The influence of the fossil fuel industry on federal policy making is legendary. It results in billions of dollars of direct public subsidies to oil companies, even though they are among the biggest and most profitable corporations in the country. Thus, it may be that the fossil fuel sector is influencing the government's projections of the nation's energy mix to suggest that renewables will remain small and unscalable throughout much of the twenty-first century. Compare the Energy Information Administration's projection of coal-fired electricity with that of Deutsche Bank, which is that the share of electricity that derives from coal will plummet from about 47 percent today to 22 percent by 2030. If Deutsche Bank and other major global investment houses are correct in predicting sharp declines in fossil fuel shares, then renewable energy and other green jobs will increase well beyond the four million cited by the Bureau of Labor Statistics.

The U.S. government is on the whole conservative in its projections of the continuing dominance of fossil fuels over renewables, but Jeremy Rifkin is an articulate spokesman for the radical energy revolution that is upon us. In *The Third Industrial Revolution: How Lateral Power Is Transforming Energy, the Economy, and the World* (2011), he contends that the information-based Internet can be revolutionized to become energy-based. In other words, just as we can instantaneously be directed to the best route to servers around the world to access desired information, in the near future, he contends, we will be able to have energy routed to different parts of the country, if not to different parts of the world, through an "Energy Internet." For this to happen, according to Rifkin, five pillars of infrastructure must come together:

1. The shift to renewable energy
2. The transformation of building stock into micropower plants to collect renewable energy
3. The deployment of technologies for storing intermittent energies (i.e., wind, solar)

4. The transformation of the mechanical power grid into a smart energy–sharing intergrid
5. The transition from fossil-fueled vehicles to hydrogen cell– and renewables-fueled vehicles

Green markets are continuing to develop in part because, many companies believe that green is no longer about just doing what is good for the environment (Deitche 2010). Markets are opening up to companies that are becoming more resource efficient and less polluting as a way to reduce cost and improve the bottom line. In addition to the expansion of markets for enhanced green production, there is the more discernible expansion of consumer markets for green products. From more fuel-efficient vehicles to organic vegetables to recyclables, companies are responding to market signals for green product lines. They are also responding to the market because more than three-quarters of Americans aged twenty-five and under say that a company's impact on the environment is important to their buying habits (Goleman 2009). Moreover, Gilles Grolleau, Naoufel Mzoughi, and Sanja Pekovic (2012) report that environment-related standards have an impact on the recruitment of high-caliber employees.

What we see is that green employment is systemic. Environmental changes concentrate the public's mind, formal and popular education prepares the public for action, governments respond to pressure for sustainability policy, and companies respond to the market's desire for clean technology. As noted previously, companies do not like abandoning business models and production processes that are efficient, reliable, and profitable. They are likely to do so only if forced by the government and/or persuaded because the new models and processes are even more efficient, reliable, and profitable.

The market may eventually propel companies to more rapidly become green as the technology databases outlined by Daniel Goleman (2009) and others come to fruition. When consumers are armed with radical product and production system transparency so that companies are not the only ones who can see their environmental impacts throughout the global supply chain, then consumers can be more assertive in punishing companies whose products and processes harm the environment and in rewarding those that do not, or at least do less so. For this to happen there must be universally accessible ecological accounting systems for determining the affects of industrial ecosystems on natural ecosystems. Such systems could make their developers wealthy and empower the public to put pressure on environmentally destructive companies. Goodguide.com was an early prototype of the dynamically evolving worldwide product database that enables consumers everywhere to evaluate products on their environmental impacts and social justice measures. In the future, as these comparative ecological accounting systems become more sophisticated, they will need the capability to prevent companies from gaming the system to their advantage at the expense of the public.

Green jobs will continue to increase, but that growth will likely neither be as fast as the decline of older, pre-green jobs nor be in the same geographical or economic sectors where brown jobs are lost first and/or more intensely. The new green jobs will require varying levels of skills, and African Americans will have to focus on the driving factors behind their increase, where they are occurring, what credentials and skills they require, and how best to compete for them. Younger students, particularly those in Africana studies courses, should keep an eye not just on the cutting edge but also on jobs in next-era clean technologies that are being funded, developed, and commercialized by ARPA-E and green venture capital, for example, and those cited by forecasters such as Rifkin or visualized by other transformative signals.

For African American and other students interested in careers such as environmental consulting, green Africana studies courses will provide sophisticated historical and politico-economic analysis of racial discrimination as it relates to technology and big-picture public policy formulation and analysis. Environmental consultants make recommendations to businesses and public policy makers on environmental issues to promote minimization of energy use and waste and enhanced sustainability. Green economic restructuring is a substantial management challenge, requiring new levels of awareness, new perspectives, and deft managerial capacities. To qualify for such a position, students must first become conversant on technical issues. For example, they will need to understand such concepts as energy return on energy invested (EROI). To harness scaled and reliable energy to meet society's needs, energy first has to be invested. Obviously the highest EROI possible is the goal, and some energy sources have higher EROIs than others. Oil sands, the dirtiest of all energy sources, have among the lowest EROIs, with a ratio as small as 3:1, which means that one unit of energy is required just to get three units out. Conversely, the average EROI of conventional U.S. crude oil is about 1:12 (Murphy and Hall 2011). Perhaps the ultimate EROI will come from the sun. In just five days, the Earth receives more usable solar energy than from all known fossil fuels. The EPA's Re-Powering America's Land initiative reports that between 2010 and 2011, jobs in the solar sector grew ten times faster than the rest of the economy. Again, spending a small fraction of fossil fuel monies to harness solar energy represents the highest possible EROI.

But Wait a Minute—Just What Exactly Are Green Jobs, and Where Are They Located?

The term "green jobs" is often used loosely, but according to the Bureau of Labor Statistics (BLS):

> The green jobs definition contains two components, an output-based approach and a process-based approach. Output-based jobs are jobs associated with producing goods or providing services that benefit the environment or conserve natural resources. Process-based jobs are jobs in which workers' duties involve making their establishment's production processes more environmentally friendly or use fewer natural resources. (Bureau of Labor Statistics 2012: 4)

An occupation, on the other hand, is a "grouping of jobs which have a common set of main tasks and duties in [a specific] sector" (Strietska-Ilina et al. 2011: 95). Green technology enhances natural resource and energy efficiency in production; reduces waste and pollution; utilizes renewable resources; and produces biodegradable, nontoxic, and recyclable resources. The UNEP defines a green economy as one that results in

> improved human well-being and social equity, while significantly reducing environmental risks and ecological sacrifice. . . . [It] is characterized by substantially increased investments in economic sectors that build on and enhance the Earth's natural capital or reduce ecological scarcities and environmental risks. These sectors include renewable energy, low-carbon transport, energy-efficient buildings, clean technologies, improved waste management, improved freshwater provision, sustainable agriculture and forest management, and sustainable fisheries. These investments are driven or supported by national policy reforms and the development of international policy and market infrastructure. (2010: 3)

As the economy becomes more green with the broader application of green technology, it will create a multiplier effect that increases green jobs. But in the course of the transformation, new production processes, products, and markets inevitably replace those already established. This progression toward green obviously means that there will continue to be new opportunities for those with the necessary skills; those without updated skills and who are in nongreen occupations will likely face joblessness.

Table 4.1 lists the major green job categories and their definitions, as provided by the BLS.

If we are to avoid environmental catastrophe, job opportunities arising from a green economy must offset pre-green job losses in absolute terms, but the offsets will not occur uniformly in geographical areas, economic sectors, or occupational categories. Those who get the new green jobs will not necessarily be the ones who lose their jobs because of greening. For this reason, I believe that historically disadvantaged groups need targeted assistance to prevent the racial and class inequalities seen in the pre-green industrial era. I also believe that green Africana studies can play a unique and indispensable role in helping the black community to better prepare for new green jobs. Assistance can come from classroom experiences as well as through outreach and urban extension programs and collaboration with local community-based organizations and community development corporations (CDCs).

TABLE 4.1 BUREAU OF LABOR STATISTICS GREEN JOBS DEFINITIONS

Category	Definition
Renewable energy	Electricity, heat, or fuel generated from renewable sources, including wind, biomass, geothermal, solar, ocean, hydropower, and landfill gas and municipal solid waste
Energy efficiency	Energy-efficient equipment, appliances, buildings, and vehicles; products and services that improve energy efficiency of buildings and efficiency of energy storage and distribution (e.g., smart grid technologies)
Pollution, greenhouse gas reduction, recycling	Pollution reduction and removal, greenhouse gas reduction, and recycling and reuse, including products and services that: • Reduce or eliminate generation or release of pollutants or toxic compounds, or remove pollutants or hazardous waste from the environment • Reduce greenhouse gas emissions through methods other than renewable energy generation and energy efficiency (e.g., electricity generated from nuclear sources) • Reduce or eliminate generation of waste materials; collect, reuse, remanufacture, recycle, or compost waste materials or wastewater
Natural resources conservation	Products and services related to organic agriculture and sustainable forestry; land management; soil, water, or wildlife conservation; and storm water management
Environmental compliance, education, public awareness	Products and services that: • Enforce environmental regulations • Provide education and training related to green technologies and practices • Increase public awareness of environmental issues

To its credit, the Department of Labor has created a much-needed office on Green Jobs for Women. It reports:

> The Women's Bureau is taking the lead in ensuring that women of all ages and socioeconomic groups are aware of and prepared to succeed in the emerging "green" jobs sector, which according to Secretary Solis will be a key driver of America's economic recovery and sustained economic stability. The Women's Bureau is collaborating with employers, unions, education and training providers, green industry organizations, and other government agencies to raise awareness, expand training options, and

promote the recruitment and retention of women in green career pathways. (U.S. Department of Labor 2012)

A similar bureau is needed for African Americans, Latinos, and other disadvantaged groups to help them succeed with the greening economy as well. Women in general are much better positioned and informed and participate more in national policy discussions on green jobs than do African Americans in general.

On the whole, we see few African Americans in government agencies with a green jobs focus. There are even fewer in the private sector. Table 4.2 shows the percentage distribution of blacks at the four federal agencies with an environment orientation as a share of the total and as a share of senior personnel, executives, scientists, and engineers.

"In 1992, whites made up nearly 74 percent of the total EPA workforce of 18,599—and . . . 95 percent of the executive ranks and 84 percent of the more senior GS ranks (levels GS-12 through GS-15)" (Rhodes 2003: 94). In 2006, whites constituted 71 percent of the total EPA workforce of 18,240. Their share of the more senior ranks was down from 84 percent to 73 percent and their share of the executive ranks was down from 95 percent to 85 percent. African Americans' share of the 2006 EPA workforce stood at 17.5 percent. Among senior GS ranks African Americans made up 15.3 percent of the EPA total and 7.5 percent of its executives (U.S. Office of Personnel Management 2006). Among all EPA employees, blacks are slightly overrepresented and have slowly increased their presence in other agencies in close touch with the environment—although they continue to be underrepresented at the Agriculture and Energy departments and especially the Department of the Interior. And they continue to be underrepresented

TABLE 4.2 BLACKS IN FEDERAL AGENCIES AND DEPARTMENTS MOST CLOSELY ATTACHED TO THE ENVIRONMENT

Agency	Total employees (%)	Senior professionals (GS 12–15) (%)	Executives (%)	Scientists and engineers (%)
EPA	17.5	15.3	7.5	8.6
Agriculture	10.8	10.1	8.6	4.9
Energy	11.1	10.3	4.9	5.9
Interior	5.1	4.3	5.8	2.2

as scientists and engineers in these departments. Of course, blacks are woefully underrepresented as graduates of programs in the natural sciences and engineering. They also continue to be underrepresented in the executive ranks, where decisions are made on the creation of bureaus for disadvantaged groups similar to the Department of Labor's Women's Bureau. According to Elsie Harper-Anderson, "To insure that African Americans and their communities are part of the new green economy, their issues and their challenges have to be an integral part of the green plan, not an afterthought" (2012: 172).

EPA region 5 has the most green jobs (543,603), followed by regions 4, 9, and 3, which range from 446,357 to 410,924. Figure 4.2 shows the distribution of Africana studies programs by EPA region. The top four regions with Africana studies programs, 9, 2, 5, and 4, also include three of the top four in number of green jobs: 5, 4, and 9. The figure also shows the number of Africana studies programs and the number of green jobs in each EPA region. Region 2, with the second-highest concentration of Africana studies programs, is sixth in number of green jobs, at least according to the Bureau of Labor Statistics (2012). Among the regions that are the weakest in green jobs are 8, 7, 1, and 10. The regions with the smallest number of Africana

Figure 4.2 Green jobs/Africana studies programs by EPA region

studies programs are 8, 10, 6, and 7. Three of the bottom four overlap. There is a striking correlation between green jobs and Africana studies programs by EPA region. Table 4.3 lists the number of green jobs in states with Africana studies programs and reveals that California and New York have the greatest number of green jobs as well as the greatest number of Africana studies programs. In fact, there is a fairly tight correlation between locations with the most green jobs and locations of Africana studies programs, not only by EPA region but also by state.

Table 4.4 lists the top ten states with the greatest number of Africana studies programs and ranks them by number of green jobs. We see that six of the top ten states for green jobs are also the top ten states for Africana studies programs. Closely related, Table 4.5 shows that six of the top ten states for Africana studies programs are also in the top ten for green jobs. This correlation was not inevitable, and it is not perfect. For instance, although Texas has the third-highest number of green jobs in the nation, it is only number 28 in Africana studies programs. Likewise, Washington is ranked 8 for green jobs but only 29 for Africana studies programs. Although New Jersey is tied with Michigan at seventh place for Africana studies programs, it is only 15 for green jobs.

Table 4.6 lists the number of green jobs by all private-sector industries. Manufacturing has the largest number at 461,847, followed by construction at 372,077. Professional, scientific, and technical services are next at 349,024, followed by administrative and waste services at 319,915. Transportation and warehousing have 245,057 jobs, followed by trade at 202,370. Utilities and natural resources and mining are respectively 65,664 and 65,050.

The Bureau of Labor Statistics reports that there were 3,129,112 green jobs in 2010 out of the nation's total of 127,820,442 jobs. Only 2.1 percent of all jobs are green, while the 860,300 green public-sector jobs represent nearly 4 percent of all jobs in the public sector. Since public-sector jobs are nearly twice as likely as private-sector jobs to be green, reductions in them are likely to reduce green opportunities, and jobs in the public sector have been cut back significantly in the past few years. For example, "between 2007 (before the recession) and 2011, state and local governments shed about 765,000 jobs" (Cooper, Gable, and Austin 2012). The only way for such reductions not

TABLE 4.3 GREEN JOBS IN AFRICANA STUDIES STATES

State	Number of jobs
Alabama	28,863
Arizona	28,815
Arkansas	26,479
California	230,758
Colorado	54,453
Connecticut	31,782
Delaware	7,033
District of Columbia	12,021
Florida	81,954
Georgia	64,810
Hawaii	11,463
Idaho	13,826
Illinois	103,344
Indiana	57,467
Iowa	32,075
Kansas	23,059
Kentucky	24,925
Louisiana	26,816
Maine	11,184
Maryland	55,742
Massachusetts	64,462
Michigan	64,615
Minnesota	56,921
Mississippi	13,551
Missouri	41,226
Montana	8,031
Nebraska	13,016
Nevada	12,361
New Hampshire	10,319
New Jersey	52,878
New Mexico	13,016
New York	134,065
North Carolina	60,574
Ohio	97,027
Oklahoma	15,694
Oregon	37,642
Pennsylvania	149,377
Rhode Island	8,804
South Carolina	29,505
Tennessee	46,205
Texas	169,367
Virginia	60,218
Washington	68,341
Wisconsin	54,771
Wyoming	4,197

TABLE 4.4 STATES WITH THE MOST AFRICANA STUDIES PROGRAMS AND RANKING BY GREEN JOBS

State	Number of Africana studies programs	Ranking by green jobs
California	60	1
New York	58	2
Massachusetts	17	13
Ohio	15	6
Pennsylvania	14	4
Illinois	12	5
Florida	11	7
Georgia	10	14
North Carolina	8	12
Michigan	7	12
New Jersey	7	15

TABLE 4.5 STATES WITH THE MOST GREEN JOBS AND RANKING BY AFRICANA STUDIES PROGRAMS

State	Number of green jobs	Ranking by number of Africana studies programs
California	338,400	1
New York	248,500	2
Texas	229,700	28
Pennsylvania	182,200	5
Illinois	139,800	6
Ohio	126,900	4
Florida	95,963	7
Washington	91,906	29
Virginia	91,871	14
Maryland	87,408	20

to constitute a net loss of total green jobs is for the private sector to increase its green jobs at a faster rate. Moreover, public-sector job reductions will probably result in fewer jobs for African Americans, since they disproportionately work in the public sector. African Americans held 20 percent of the 765,000 public-sector jobs shed between 2007 and 2011.

Figure 4.3 shows the correlation between concentrations of brownfields, toxics release inventory (TRI) facilities, green jobs, and Africana studies programs by EPA region. Its most prominent takeaways

TABLE 4.6 GREEN EMPLOYMENT IN PRIVATE INDUSTRY, 2010

Industry	Number of jobs
Manufacturing	461,847
Construction	372,077
Professional, scientific, and technical services	349,024
Administrative and waste services	319,915
Transportation and warehousing	245,057
Trade	202,370
Utilities	65,664
Natural resources and mining	65,050
Information	37,136
Education and health services	37,069
Management of companies and enterprises	34,711
Leisure and hospitality	22,510
Financial activities	190
Other services except public administration	56,174
Total	**2,268,824**

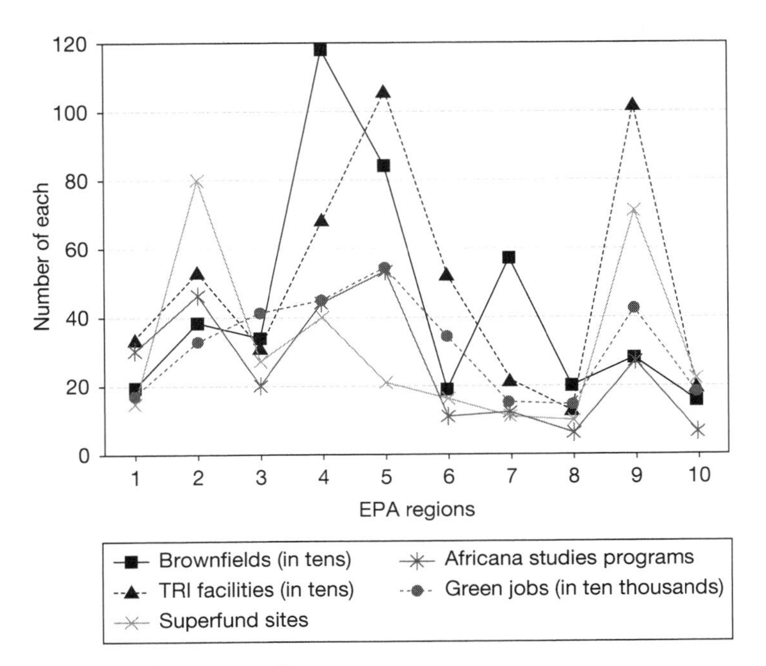

Figure 4.3 EPA region totals

are that there are considerable opportunities for Africana studies programs around the country to engage with black communities on the environment. More broadly, in EPA regions and in states where Africana programs exist, green jobs already exist in sizable numbers that are expected to grow. As the figure suggests, more Africana studies programs exist in EPA regions and states that have more brownfields needing remediation and redevelopment, more TRI facilities, and more green jobs. All of this means that, as noted in Chapter 1, there is much room for Africana programs to prepare students for the environmental needs of the black community as well as for current and future green careers.

In addition to the private and public sectors, environmental organizations, by definition green, offer jobs in the nonprofit sector. Mainstream environmental organizations either have historically privileged preservation and conservation over immediate environmental health concerns of working-class communities and communities of color or have worked to restrict toxins and the destruction of ecosystems as if such threats were universally distributed. Environmental justice organizations, which are not generally considered mainstream, focus on the overconcentration of environmental burdens carried by communities of color and working-class communities. However, mainstream organizations generally are better financed than environmental justice organizations, which means that they can offer more opportunities for employment.

One problem is that mainstream environmental organizations have been slow to diversify their staff and leadership. Dorceta Taylor (2007a) has done pioneering work at the University of Michigan in documenting the paucity of communities of color in environmental jobs in all sectors. She has engaged such institutions on their lack of diversity and their failure to prepare individuals from those communities for professional positions. Two decades ago, according to Taylor, less than 2 percent of workers in the National Audubon Society, the Sierra Club, Friends of the Earth, and the Natural Resources Defense Council were minorities (Taylor 2008). She also notes that in a study by the Environmental Careers Organization, among sixty-three mainstream environmental organizations, 32 percent had only white employees on staff in 1992. By the 2000s mainstream organizations had come to acknowledge the diversity problem and had begun to address it, but only after charging that black and other communities of color had little interest in and lacked the preparation for employment. In 2000, approximately 11 percent of staff members in approximately sixty environmental organizations affiliated with the Natural Resources Council of America were from communities of color (Taylor 2008).

To gauge and compare the outlook, expectations, and desires of blacks and other students for working in environmental organizations, Taylor administered five thousand surveys to those who were in the science and engineering programs most closely related to the environment at 185 colleges and universities. Programs of study included agricultural sciences, environmental engineering, environmental science, social sciences, natural resources, biological sciences, forestry, geography, and geosciences. Overall, students showed the greatest interest in government agencies and academia and showed the least interest in environmental justice organizations (Taylor 2007b). Taylor reported that "blacks (63.2 percent) were far less likely than others to say they were willing to work for mainstream environmental organizations. Whites (49.7 percent) were the least likely to indicate a willingness to work for environmental justice organizations." Further, "blacks (92.6 percent) were far more likely than others to say they were willing to work in nonenvironmental government agencies. Blacks were also most likely to say they were willing to work in corporations upon graduation" (2007b: 179). In Taylor's study (2007b) 37.2 percent of white students regarded a job in academia as ideal, and 37.0 percent of black students regarded a job in a government environmental agency as ideal. Taylor's empirical research debunked mainstream environmental organizations' contention that black and other communities of color are rarely hired because they lack academic preparation, lack interest in such employment, and/or have salary expectations grossly out of line with actual compensation.

The point of this chapter is that not only are Africana studies students in America, but America is in them, too. Some aspire to status-laden and materially rewarding careers. But some are drawn toward Africana studies for its social mission, its alternate way of understanding society, and its unique opportunities for self-discovery. Green Africana studies allows all of these aspirations to be fulfilled.

Africana studies and other liberal arts fields have an important role in the global transformation to a green economy. Thus, they are at least as important as the STEM disciplines in this transformation, which will be one of the key milestones in human experience. The destructo-industrial economy created tens of millions of jobs and made it possible for worldwide urban-based human needs to be met, but the price extracted was a destabilized environment. The global green economy will also create tens of millions of new jobs and at the same time make it possible for all to have access to limitless clean energy, nontoxic products, and efficient and renewable production that will prevent the exhaustion and weakening of ecological systems.

While important green jobs may be available for low-skilled workers in urban areas, it is unlikely that there will be enough for all who need or want them. Transnational capital and the fossil fuel industry, as well as the political forces that represent them in Washington and in the fifty state capitals, will continue to block public investment in greening the urban core in their short-term interest and to the extent of their power. To some degree, the number of green jobs will reflect not environmental or job needs but rather the outcomes of ongoing political struggles between pro- and anti-green forces in the American political economy. Green Africana studies provides students with an excellent education on these points, thereby preparing them for potential leadership positions and girding them for victory on the battleground of ideas

Greenwashing

Some analysts are much more pessimistic than I am about the prospect of significant numbers of green jobs in the future. Considering the short term, however, I agree with the staunch pessimists that the abundance of green jobs that advocates have written about since 2000 is unlikely to occur. The prospects for green jobs that I have suggested are not based on my acceptance of uncritical views of "greenwashing" as a sales job. The term "greenwashing" has been defined as "the intersection of . . . poor environmental performance and positive communication about environmental performance" (Delmas and Cuerel Burbano 2011: 65). This concept inflates the outlook for green jobs because what it amounts to is a marketing spin declaring that investments to improve environmental performance and increase environmental benefits through new products and services are really going to happen. In other words, if we buy into the hype about all of the green production supposedly forthcoming—that is, believe in the transformation of brown into green industry and that government funds for the green economy will be spent—then we must conclude that an abundance of green jobs is indeed impending.

The problem is that there is often a wide gap between positive symbolic communication about positive environmental performance and substantive action by industry and the government that is required to achieve it. When all the posturing is cleared away, the requisite public and private spending and political will are clearly seen to be woefully inadequate for creating the large number of future green jobs in the intermediate term being bandied about by advocates.

Greenwashing is not only about companies' promotion of their green performance and government hype about public investments; it is also about downplaying the environmental degradation that companies and governments generate and seek to continue. Arguing that fossil fuel–led anthropogenic climate change is not real is a consummate example of greenwashing discouragement, which aims to either lessen the perceived seriousness of environmental damage or argue for more study of the problem. Both greenwashing promotion by industry and government and greenwashing discouragement by industry, government, and front groups conceal the true number of potential green jobs. This sets limits on green jobs by damping consumer pressure for green products, services, and production methods that might come about much faster and in more revolutionary ways than polluting industries want.

Currently, the American public has little access to serious reporting about environmental problems and opportunities. Even when highly credentialed members of the scientific community present overwhelming evidence that anthropogenic climate change is the gravest threat to humanity, news organizations, in the alleged interest of equal time, provide opportunities for the "other side" to offer counterarguments that it claims also have a scientific basis. This type of reporting muddies the water and confuses the public, which again reduces the pressure on industry and the government to act on behalf of the environment and ultimately on behalf of humanity. According to Sharon Beder, "Most media organizations are owned by multinational multi[billion-]dollar corporations that are involved in a number of businesses apart from the media, such as forestry, pulp and paper mills, defense, real estate, oil wells, agriculture, steel production, railways, water and power utilities. Such conglomerates not only create potential conflicts of interest in reporting the news, but also ensure that the makers of the news take a corporate view" (1998: 220).

Notwithstanding such greenwashing, the public still wants more renewable energy investments and more environmental protection. Ironically for the public, greenwashing promotion and greenwashing discouragement lower the prospects for green jobs during the intermediate period because the campaign carried out by the media, industry, and government has had great success in suppressing public pressure.

A point lost on neither state-centric theorists nor capitalist class–dominated state theorists is that economy-wide replacement of an energy source does not simply mean a new industry replacing an older one. Over the past few centuries, the world's dominant economic power at any given time was associated with a prevailing energy source, and when that source was replaced by its successor, dominant economic power was replaced as well (Patterson 2010). For example, "the Dutch rose to world economic leadership through genius with wind and water and could not maintain commercial hegemony under a new international fuel regime. Coal did the same thing for Britain, until the rise of oil-powered industry and military forces gave the edge to the oil-favored United States" (Phillips 2006: xlvi). U.S. corporations and political leaders are no doubt aware of this history, and the operating logic—that advanced renewables will replace oil, opening a pathway for some other hegemon to replace the current one—still holds.

Greenwashing is another way of saying that industrial, financial, and political elites are not making the mitigation of environmental destruction and the pursuit of a green economy their highest priorities. As long as this is the case, then, yes, the predictions about green job opportunities given earlier may in the intermediate term prove too optimistic. After all, companies are loathe to abandon profitable business models. They do so only when forced by government regulation or when there is a strategic shift regarding the acceptable risk of a new business model that is at least as profitable as the existing one, likely for a longer duration. The historical "brown business models" appear, on the whole, more profitable than "green business models," because they are in part the result of built-in government subsidies. As long as those subsidies continue, green jobs will be limited.

I am nevertheless bullish on the significant availability of green jobs in the long term because, if we are to avoid future environmental collapse and a human catastrophe, there will have to be substantial green production and sustainable consumption, which will require an abundance of green workers. The world's population is growing by more than ninety million every year, and that growth is almost totally in the developing world, whose inhabitants have long dreamed of emulating the lifestyles of peoples of the developed world—the Global North—with their motorized vehicles, central heating and cooling, and electric appliances, but now these underserved peoples are rapidly making material progress to this end. The United Nations Development Programme (2013) projects that by 2030, 80 percent of the world's middle class will be in developing nations, once known as the third world. The only way that more people can continue to materially improve their lives without accelerating the deterioration of the planet is for eco-industrial technology and associated green jobs to be available on a global level. The mission and proposed collaboration of Africana studies and environmental studies can be instrumental in such an achievement. In the end, if some are pessimistic about green jobs in the long term, they are also pessimistic about the prospect of humanity avoiding environmental collapse.

At this point I have made a case for the intellectual and professional integration of Africana and environmental studies as a means for each to fulfill its laudable mission. The focus thus far has been on "African America," or the African American experience, to some extent reflecting the fact that Africana studies focuses most of its attention there. Each of the hundreds of Africana studies programs in the United States calibrates its emphasis on African America, Africa, and other parts of the diaspora to reflect its own expertise, interests, and campus experiences, but the African American experience is privileged because it is the preference of African American students and administrators at predominantly white universities and colleges.

That said, any credible Africana studies program must provide proper coverage of Africa, both historically and contemporaneously. In the 1960s and 1970s when the push for Africana studies programs was at its strongest, Africa loomed large in the African American imagination as African nations were gaining their independence and becoming more important on the world stage. Now we are on the cusp of an urgency to connect with Africa as a scholarly subject and to engage the continent professionally as it continues its ascendance in the global economy.

To engage with Africa fully, public, private, and nonprofit institutions in the United States and Africa will require the professional services of many experts on African nations. Africana studies programs offer the best training for developing such expertise, because they alone possess the historical perspective, the sensitivity, and the affective connection to best serve Africans first and foremost and the institutions that seek to engage them. Moreover, green Africana studies will provide students with sufficient training to address one of the continent's greatest challenges and one of its greatest opportunities—the environment.

The next chapter covers Africa, which is experiencing economically transformative changes at an unprecedented pace in this modern era.

VIGNETTE 4.1. JILL HUMPHRIES

Seeing the World through Nature's Eyes

If you love to travel, be with nature, and learn about world culture, then think about pursuing a career in ecotourism. My interest in ecotourism grew out of my global social justice work in Southern Africa over twenty-five years ago. Although I was focused on political issues, one cannot travel to Africa and not be awed by the sheer size and nature of its landscape. Being a city girl, I had never really experienced nature on its own terms. I remember living in a rural area—Old Mutare, Zimbabwe, nestled in the northeastern corner of the country. I could see the Mozambican mountains, the Inyanga pine forest (my favorite spot), and the Matopos' free-standing rocks, which are the size of buildings.

There were creepy-crawler spiders on my bedroom wall, but I learned that they were my friends because they ate the other creepy-crawlers hiding out in my room. My Zimbabwean students taught me the significance of respecting nature in the forested area that surrounded our mission station—including the stealthy mambo green snakes that drop out of trees onto your head and can kill you with just one strike. But that didn't prevent me from hanging out with the local Shona artist, who introduced me to mountain climbing and camping trips. My most intriguing experiences, however, were with the N'angas, the traditional healers of Zimbabwe. I interviewed them about their practices and how they lived with nature and of course solicited a reading for myself! They responded, "While you here, you are one of us." These experiences forever shaped my relationship to nature and culture and my understanding of the importance of global ecostewardship. They also prompted me to establish my own ecotourism business, RUDO Ecotravels, in order to provide these opportunities to communities of color.

It is for these reasons and many more that I believe the ecotourism sector offers exciting green employment opportunities for young African Americans. The shift toward responsible and sustainable tourism over the past thirty years has prompted growth in ecotourism, a subsector within the tourism and travel industry. "Tourism and travel" is considered the largest employment sector in the world economy. According to the World Travel and Tourism Council's 2011–2012 annual report (2012):

- *Tourism contributed almost $6 trillion to the global economy, or 9 percent of global gross domestic product (GDP).*
- *The industry employed nearly 260 million worldwide in travel and tourism and related sectors.*
- *Global tourism GDP is projected to grow by 54 percent over the next decade—that is, to $2,860 billion by 2021.*
- *The travel and tourism industry is expected to create an additional 21 million (net) direct jobs over the next decade.*
- *Ecotourism is the fastest-growing sector of the tourism industry.*

There is no consensus within the field on how to define ecotourism; however, the International Ecotourism Society (TIES) offers a useful definition that is universally accepted: "Ecotourism is responsible

TABLE 4.7 TOURISM TERMINOLOGY

Activities	Impacts
Tourism: Travel for recreation	Ecotourism: Responsible travel to natural areas that conserves the environment and improves the welfare of local people
Nature tourism: Travel to unspoiled places to experience and enjoy nature	Responsible tourism: Travel that maximizes benefits to local communities, minimizes negative social or environmental impacts, and helps local people conserve fragile cultures and habitats
Mass tourism: Large-scale, typically associated with "sea, sand, sun" resorts, transnational ownership, minimal direct economic benefit to destination communities, seasonality, and package tours	Sustainable tourism: Travel that meets the needs of present tourists and host regions while protecting and enhancing opportunities for future travelers and the region
Adventure tourism: Nature travel that involves physical skills, endurance, and risk taking	Geotourism: Travel that sustains or enhances the geographical character of a place—its environment, heritage, aesthetics, and culture—and the well-being of its residents
Cultural tourism: Travel for essentially cultural motivations	Pro-poor tourism: Travel that results in increased net benefit for poor people

Note: Adapted from Center for Responsible Travel, n.d.

travel to natural areas that conserves the environment and improves the well-being of local people" (n.d.). The Center for Responsible Travel's tourism terminology (Table 4.7) highlights the shift from activity to impact-based tourism as the major defining feature of ecotourism. This shift over the past two decades to value-based tourism is based on the principles and practices of sustainable development.

Ecotourism Actors

The ecotourism sector comprises a variety of actors, including government (federal, state, city, and local communities); higher-education institutions; foreign consulates; private-sector actors such as tour operators and guides, hotel builders, and marketing officials; subsidiary industries such as local businesses, restaurants, and souvenir stores; and cultural, community, and environmental centers.

Government actors are responsible for establishing an enabling environment for the tourism and travel industry. This includes:

- *Passing the Federal Promotion Act to create a nonprofit corporation*
- *Protecting national parks and biodiversity sites*
- *Protecting antiquities and cultural historical sites*
- *Promoting job creation in the tourist industry*

How Should I Prepare for a Career in Ecotourism?

There are many employment paths within the ecotourism sector that you can pursue based on your interests, education, and skills. If you wish to pursue nature-based ecotourism, becoming an expert in ecosystems by pursuing a B.A. degree in natural wildlife and resource management or environmental sciences is one approach. If you are more interested in people, places, and culture, a degree in anthropology, Africana studies, ethnic studies, human geography, or history is more appropriate. However, the best approach is to pursue a dual major/minor that explores the human-nature nexus. On the basis of my travels, I find that the most successful ecotourism businesses employ individuals who are experts on nature and culture. Pursuing a formal education is one such way to acquire this expertise, as well as technical expertise, language competency, statistical analysis, and mapping. All course content should be supplemented with applied learning such as internships and on-the-job training in both a domestic and a global context. Also take advantage of opportunities to study abroad through your Africana studies department and other campus programs.

African American young adults might also be drawn to ecotourism because of its core values and focus on justice and equity issues in human-environment relationships. This environmental justice framing provides the context in which to examine and develop skills in ecotourism planning and community development.

A park ranger is one example of an ecotourism worker. Park rangers are employed at every level of government and are responsible for protecting our national, state, and city parks. As park stewards, they are also responsible for environmental education and historical interpretation. As a New York City Urban Park Ranger Summer Fellow (2008), I learned firsthand how rewarding it is to work with youth of color as they learn about the natural world and Native American history.

Historical Interpretation: The African Burial Ground National Monument

As previously discussed, New York City parks are sites of both ecocultural and historical points of interest. The African Burial Ground National Monument, located in lower Manhattan, is the largest colonial African burial ground in North America. It was designated as a national monument in 2006. The project began in 1991 during excavation for a construction site, where skeletal remains of over four hundred men, women, and children from the seventeenth and eighteenth centuries were found. Further research revealed that free and enslaved Africans were buried in this 6.6 acre burial site, which was outside the colony of New Amsterdam (see http://www.nps.gov/afbg/index.htm).

The discovery of this magnificent archaeological jewel has caused a major rethinking of colonial historiography, African/black Atlantic studies, and, in particular, the significance of the African influence in the Americas. The National Park Service has done a wonderful job in creating an interpretive center at the site and in providing several online reports for public access. In addition, the center has historical interpreters who lead daily tours of the burial ground, informing visitors of the life and times of African people in colonial America. I encourage students to visit our country's local, state, and national parks given the rich history of these sites and their value in educating the general public about the significant contributions of our multiethnic past.

For many youth and young adults of color their first ecotourism experiences are with urban parks within the metropolis. This provides a great opportunity to prepare them for working in the ecotourism sector. Perhaps drawing from our national parks as a model, we will come to see the beauty and significance of city parks and invest more money and human capital in promoting them as both natural and cultural heritage sites. Promoting the multicultural history of these sites is one way to generate interest among African American young adults in becoming park rangers. As one can see, New York City's urban parks offer much to the city's landscape!

5/ Greening and Growing Africa Economically

A Role for Transnationalism

Africa's Economic Transformation and Environmental Challenges

Finally! Afro-pessimism has given way to Afro-optimism! For example, Africa is experiencing declines in child mortality faster than the world has witnessed anywhere for the last thirty years ("The Best Story" 2012). Since the late 1990s, it has been on a continuous economic growth trajectory, and the fundamentals make it such that this is not the false dawn witnessed in years past. This time, economic growth and transformation are represented not just by an expansion of commodities in response to churning development in wealthy core nations or the emerging giant of China; rather, they are being seen in telecommunications, retail, manufacturing, construction, banking, transportation, and in other arenas. This pessimism-to-optimism shift is reflected in the hugely contrasting headline stories that appeared in two separate editions of the *Economist* over an eleven-year period: a May 2000 cover story calls the continent "Hopeless Africa," but a December 2011 cover story screams, "The Hopeful Continent: Africa Rising." The latter article certainly appears to be a mea culpa for the pessimistic headlines of over a decade ago—the magazine now gets the story right. Notwithstanding the Ebola crisis that has killed more than two thousand citizens in few West African nations in 2014, the continent is positioned to progressively transform economically and socially. This chapter is about the economic transformation under way in Africa, the necessity of economic growth quickly becoming much greener, and the role of transnationalism and the African diaspora in achieving these goals.

Most people who do not follow the twists and turns of Africa would be surprised to learn that seven of the world's ten fastest-growing economies are there. Over the past fifteen years or so, Americans have no doubt been largely oblivious to the continent's political, economic, institutional, social, and cultural changes, of which "the biggest . . . is economic," according to Lydia Polgreen (2012). She observes, "The growth rate for the continent has crept up, rivaling Asia's overall growth at the height of the 'tiger economy' era of the 1990s, and it could reach 7 percent by 2015, according to the United Nations Development Program." Polgreen also cites the World Bank's recent conclusion that "Africa could be on the brink of an economic take-off, much like China was 30 years ago, and India 20 years ago." For example, Africa's economic growth rate in 2012 was twice as high as Brazil's, and its economy and its middle class are larger than India's. Business publications such as the *Financial Times* and the *Harvard Business Review* and consulting giants such as PricewaterhouseCoopers and McKinsey and Company have all documented Africa's economic transformation.

Table 5.1 lists seventeen of the fastest-growing African countries since the mid-1990s, their GDP-expressed purchasing power parity (PPP) in 1995–1996 and 2007, and their cumulative increase in average real income between 1996 and 2008 (CIA 1995–1996, 2007; Radelet 2010). Some have dubbed the economies of these countries as emerging (Radelet 2010). Data show, moreover, that the average income for these countries increased by 50 percent over the twelve-year period.

George Ayittey (2006), in *Africa Unchained: The Blueprint for Africa's Future*, argues that Africa's transformation is in part because of the "cheetahs" replacing the "hippos" in government, business, civil society, academia, and journalism. The cheetahs in Ayittey's metaphor are the younger generation of educated professionals who focus more on innovation, accountability, and good governance; the hippos

TABLE 5.1 GDP AND INCOME GROWTH IN THE EMERGING AFRICAN COUNTRIES

Emerging African countries	1995–1996 GDP in purchasing power parity (in billions of $)	2007 GDP in purchasing power parity (in billions of $)	Cumulative increase in average real income, 1996–2008 (%)
Botswana	4.6	26.4	68
Burkina Faso	8.0	17.4	43
Cape Verde	0.5	1.6	67
Ethiopia	24.8	56.1	65
Ghana	27.0	31.1	40
Lesotho	3.7	3.1	33
Mali	5.8	13.6	37
Mauritius	11.7	14.3	61
Mozambique	12.2	17.6	96
Namibia	6.2	10.7	36
Rwanda	3.8	8.1	60
Sao Tome and Principe	0.2	0.3	40
Seychelles	0.5	1.4	37
South Africa	227.0	468.0	29
Tanzania	18.9	51.1	46
Uganda	16.8	29.1	61
Zambia	9.7	16.1	25
Average			50

Note: Data from CIA 1997; CIA 2007; Radelet 2010.

are those who focus more on the centuries-long destruction wrought by imperialism and colonialism. Although Ayittey's formulation is interesting, it is also a bit narrow; for instance, one might consider China, with the world's fastest-growing large economy, as also run by cheetahs who nonetheless remember the exploitative imperialism and humiliating domination that their country endured. It is not necessary for knowledge of past deeds and knowledge of present needs to be mutually exclusive.

Africa's continuous growth—that the quality of life for hundreds of millions of Africans is improving and that they have increasing choices in their lives—is the good news. But there is bad news, too, as demonstrated in Chapter 3 in particular. Environmental costs come with the broader and deeper deployment of modern fossil fuel technologies. Throughout most of the current industrialization, Africa has disproportionately suffered from the adverse environmental use of these technologies while receiving few of their benefits. Clearly, Africa has begun to see more of the benefits, but the adverse environmental outcomes are still accumulating.

Africa has experienced economic growth despite being energy impoverished. For example, twenty-two of sub-Sahara's forty-nine countries have less than two hundred megawatts of generating capacity, which is enough power to provide electricity to only one hundred thousand American households—about the size of Palm Bay, Florida. Economic progress despite this limitation is indeed a reason for optimism regarding Africa's transformation.

There is, of course, an understandable relationship between energy poverty and economic poverty. Economic activity takes natural resources and adds value to them to manufacture products for local and global markets. Adding value competitively and at scale requires energy well beyond muscle power. Nations such as those in Africa that are seeking to accelerate their economic growth are largely responsible for the fact that global energy demand will increase by 50 percent in less than two decades, and they will consume 90 percent of the new energy. At the same time, Africa disproportionately suffers health-related problems resulting from present-day climate changes, which cause the death of more than three hundred thousand individuals each year (Global Humanitarian Forum 2009).

What I am suggesting is that Africa's economic growth comes with environmental and health costs associated with today's state-of-the-art technology, and this creates a conundrum: more people need modern energy, clean water, better nutrition, and more of the basics of life, but the methods to meet those needs can eliminate the improvements gained, and more, because of environmental destruction. The now famous World Commission on Environment and Development (WCED) report, known as the Brundtland Report, stressed humanity's ecological interdependence and our common stake in diminishing and ultimately reversing today's environmental damage: "Economics and ecology must be completely integrated in decision making and law making processes not just to protect the environment, but to protect and promote development" (United Nations 1987).

The 2007 *Human Development Report*, not unlike many other United Nations Environment Programme (UNEP) and United Nations Development Programme (UNDP) documents, provides a litany of important findings. For instance, approximately 98 percent of the shock effects of climate change occur in the Global South (United Nations Development Programme 2007). Additionally, one U.S. state, Texas, with a population of 23 million, has a higher carbon footprint than do the 856 million people of sub-Saharan Africa (United Nations Development Programme 2012). Sub-Saharan Africans are responsible for only about 2 percent of the world's total carbon emissions, much of them from industries in South Africa or deforestation throughout the continent. Africans are responsible for relatively few of the sources of climate shocks, but they suffer immensely from their effects.

Climate shocks are manifested as droughts in some parts of Africa and floods in other parts. An associated problem is an environmentally induced higher incidence rate of cholera and malaria, the latter from increased rainfall, temperature, and humidity. Every year, malaria snuffs out the lives of roughly nine hundred thousand individuals, nearly 90 percent of them children under five. Climate shocks also undermine agricultural production, which accounts for approximately 40 percent of Africa's GDP and nearly 60 percent of the population's livelihood (Chilonda, Machethe, and Minde 2007). Africa may have 60 percent of the world's arable, uncultivated land (McKinsey Global Institute 2010), but much smarter, more determined efforts are needed to limit its environmental degradation and protect its ecological services.

In sub-Saharan Africa, 94 percent of the rural population and 73 percent of the urban population use biomass (wood, charcoal, crop residues, animal dung) and coal as their main source of energy for cooking and heating (Ezzati 2005). Biomass burning is debilitating in many instances and deadly in others; it takes the lives of thousands of sub-Saharan African children because of respiratory infections and kills thousands of adult women because of chronic obstructive pulmonary disease per year (Ezzati 2005). Also, the deforestation to produce biomass not only accelerates global warming but also adversely affects soil, agricultural production, and other life-sustaining ecosystems and practices. Alternatives to biomass burning as a form of energy and light are urgently needed. When the sun goes down in Africa, the lights go out because more than 60 percent of sub-Saharan

Africans are not connected to an electric grid. However, if they were to receive fully fossil fuel–based electrification, their carbon footprint would skyrocket, and the climate shock to the continent would become more severe much faster. There is a palpable need to replace both global state-of-the-art high-tech electrification and Africa's low-tech biomass burning with future-tech sustainable, low-to-no-carbon electrification.

Africans and their United Nations partners, such as UNEP and UNDP, have responded to environmental shocks defensively. By that I mean that they are trying to establish conventions and regulatory regimes, extend capacity building to assure enforcement, and basically prepare to contain an assortment of problems to limit the number of individuals who perish and the number of societies that are weakened by environmental degradation. The African Ministerial Conference on the Environment (AMCEN), established to tackle environmental degradation and its attendant poverty issues, is arguably the highest-level, most authoritative, and most visible indigenous African institution concerned with the environment. My use of the term "defensive" to describe the efforts of AMCEN and its UN partners reflects my observation that a deliberate defense against environmental destruction is certainly understandable given that unchecked anthropogenic environmental changes that seriously affect rainfall, disease containment, and biodiversity are life-and-death issues.

AMCEN was established in 1985, partially in response to the Brundtland Report, to serve as Africa's principal policy forum for addressing shared environmental issues. An offensive posture, not presently a part of AMCEN's portfolio, would entail the internal transformation of African economies and their elevation in the global economy concomitant with limits on the effects of environmental degradation. AMCEN has noted that in addition to Africa facing the world's gravest environmental challenges, it also has the least institutional, human, and financial resources to deal with them (African Ministerial Conference on the Environment 2006). Given these circumstances, a green Africana studies curriculum could be of great help in preparing professionals with requisite skills and a perspective for mobilizing the continent's resources sustainably and equitably.

The AMCEN bureau, which is based at the UNEP African Region Office in Nairobi, Kenya, is composed of one member from each of the five regions of Africa: Northern, Eastern, Southern, Central, and Western. Nations in each region address conference-identified issues tailored to the region and identify issues to be sent back to the conference for prioritizing and support. Despite some limitations, such as insufficient funding and inefficient administrative structure, AMCEN has done and continues to do important and valuable work, including producing reliable, accessible, and harmonized environmental data for Africa and holding training workshops to promote environmental education. While AMCEN's ambitions do not quite measure up to the environmental and economic challenges facing Africans, they nevertheless tower over the resources provided to AMCEN by the governments.

One offensive, or forward-leaning, action for AMCEN would be to follow the African Union's lead in embracing the African diaspora as the continent's sixth region. Doing so would bring diaspora intellectuals, social networks, and other resources together to tackle Africa's environmental challenges and pursue its environmental opportunities. Africans and African Americans, as well as others of the African diaspora, suffer disproportionately from environmental degradation and climate injustice. Their struggles can both inform one another and help establish a critical mass of energy and insight for sustainable economies in their respective societies. I have been describing such a proposed African transnational project in various publications since 2005 (Patterson 2013b, 2012, 2011, 2007, 2006).

AMCEN is regarded as the implementation arm of the New Partnership for Africa's Development (NEPAD) Environment Initiative. NEPAD is a multisectoral, intergovernmental organization comprising the continent's fifty-five African states. It is charged with providing an evolving vision and strategic framework for Africa's renewal. The huge array of its initiatives and programs is designed to improve, through sustainability, the well-being of the continent's inhabitants by accelerating the eradication of poverty and the ending of inequality between the continent and advanced industrialized nations. The intent is to stop the marginalization of Africa in the global economy. Essentially, "the plan relates to Africa's common and shared sustainable development problems and concerns. It is a body of collective and individual responsibilities and actions that African countries adopt and will implement to maintain the integrity of the environment and ensure the sustainable use of their natural resources through partnerships with the international community" (New Partnership for Africa's Development 2003). The action plan is organized around clusters of activities: combating land degradation, drought, and desertification; protecting wetlands and marine and coastal resources; eliminating invasive species; ensuring cross-border conservation of natural resources; working collaboratively across Africa and with the broader international community to limit climate change; and addressing complicated cross-cutting issues.

African Transnationalism and the Environment

African transnationalism is both a field of study and a set of lived experiences through which migrants and homeland-based citizens construct, navigate, and nurture social fields that intimately and materially link the African homeland to the African diaspora. Transnational societies are those that see imagined communities straddling borders and citizens not only intimately integrated into multiple societies but also materially engaging more than one in a dynamic fashion. Governments in the Global South and international development organizations, including UNDP, have come to appreciate the special—even indispensable—role that transnationalism can play in elevating peripheral countries. Africana studies is uniquely positioned to educate and prepare students for African transnational projects, if not Pan-African projects, related to environmental sustainability.

Perhaps no development strategy has garnered more attention over the past decade than migration and its attendant aspects of homeland-diaspora engagement (Patterson 2006). A broad analysis of migration with respect to nations progressing from the periphery to the semi-periphery and even from the semi-periphery to the core has culminated in what is known as the migration-development model. This model does not rest on an elegant or even a grand theory. Instead, it is the product of an extensive body of national case studies. The evidence of the case studies is striking in support of the conclusion that national development essentially presupposes substantial contributions from the diaspora. In other words, it is almost a prerequisite that nations of the Global South first successfully engage their diasporas in core nations before they can both reposition themselves in the global economy and enhance the quality of life of their citizens.

Indeed, the migration-development model has become central to nations in the Global South that have advanced from the periphery to the semi-periphery, and its popularity has grown with African governments. Perhaps the most prominent proof of this popularity is the 2001 request of the African Union's predecessor, the Organization of African Unity (OAU), for support from the International Organization for Migration (IOM) and the establishment of the Migration for Development in Africa (MIDA) program. MIDA, as its name states, is intended to transform emigration from a brain drain into a brain gain. Key aspects of the migration-development strategy involve governments of the Global South working with their nationals in core nations to enhance their human and economic capital. The idea is that the nationals maintain economically productive relations with their host countries so that some of the acquired human capital will be invested in the homeland (Patterson 2013a).

Initially after gaining independence, African governments encouraged nationals to study in core countries since colonialists, almost unfailingly, had neglected to build modern higher-education systems in their territories. When higher-education systems were established, they were for the benefit of the children and institutions of the colonizers. The thinking among new African governments was that their nationals would "learn and return" if not "learn, earn, and return." In any case, repatriates were significantly fewer than had been hoped for, so governments began to apply the brakes to emigration. Another reason for the state turning against white-collar emigration was the challenging political agendas of those returning. Africans repatriating from the core tended to push for wider civil society space and greater transparency and accountability. The ruling elites and regimes systematically responded by attempting to curtail the number of emigrants and repatriates as well as by limiting their influence in the homeland while residing in the core. Hein de Haas (2007) describes the historical undulation of African governments' positions on emigration as a strategy for national development: first, for it, then against it, and now for it.

In many respects, both the push toward strengthening African transnational societies and the use of the migration-development model were ratified by the heads of state at a meeting of the OAU in Lusaka, Zambia, in 2001. Conceptually speaking, African governments were sold on the model at that time, if not before, but in reality their reticence, for reasons just mentioned, hampered the model's effectiveness in the African context. When African governments have created ties to their diasporas, they have tended to do so through either foreign affairs or homeland affairs ministries. But to be truly effective, as demonstrated by a number of empirical case studies, the head of state must be fully engaged in driving the migration-development agenda to ensure energized action by all ministries, certainly including foreign affairs, homeland affairs, education, finance, planning, and labor, among others.

In Asian nations that successfully applied the migration-development model, their heads of state drove several ministries to collaborate to make it work. One of the first Asian nations to apply the model was India, in the late 1940s, albeit very unsystematically at the outset. As prime minister of the newly independent country, Jawaharlal Nehru saw this approach as indispensable for strengthening the country's institutional capacities. Educated at Cambridge, he encouraged his citizens to take up studies and commercial enterprises in the core to gain economic as well as human capital. Nehru was a pioneer in migration development, but subsequent Indian heads of state were skeptical of sending to core nations large contingents of India's most educated for human and economic enhancement in the hope that India would one day benefit economically upon their return and reinvestment. Six decades after Nehru's inchoate articulation and acceptance of the migration-development model, the country is now becoming a global leader in migration-development strategy. Indian alumni from Indian Institutes of Technology are wooed by top universities in core nations for graduate study. Once credentialed they often find employment at leading technology firms, R&D centers, investment houses, and universities, amassing a wealth of technical and tacit knowledge that is mobilized for commercialization and social entrepreneurship in their homeland. The Indian example is there for Africans to follow, particularly in the area of renewables.

African governments have recognized and yielded to the inevitability of emigration in a world that is increasingly marked by the fluid movement of peoples across a single economic space. The attitude now is "if you can't beat them, join them." Since emigration is going to occur, and since other nations have utilized the migration-development model with rousing success, there is now a willingness to apply it in some African countries. The migration-development model is essentially brain circulation, a form of systematic brain gain because it meets the goals of all stakeholders: foreign students and host universities, migrant workers and their employers in the core, families of migrants, and the governments of both the host and homeland nations (Patterson 2006; Saxenian 2006). The emigration of talent out of the Global South to the West is not a brain drain when its systematic orchestration is beneficial to all stakeholders. Many Asian nations (e.g., China, Taiwan, South Korea, India) have long utilized brain circulation, and Latin American nations (e.g., Mexico, Colombia, Chile) have recently adopted it. For African nations, however, only ad hoc as opposed to systematic brain circulation has been seen (Patterson 2013a). Brain drain occurs when emigrants refuse—or are not invited—to return to their homeland or they refuse to reengage from afar via knowledge or other forms of capital transfer, because the conditions that propelled them to flee still remain.

The process of moving from systematic brain drain to ad hoc brain gain and then to systematic brain gain or brain circulation is directional but not inevitable. One cause of brain drain is intolerable living conditions in the Global South because of poor infrastructure and low human development combined with a politically repressive and culturally stultifying regime. In this case the migration-development strategy cannot work because diasporans in core nations refuse to return to or invest in one form or another in their respective homelands. Brain circulation accelerates when, in addition to the ruling regime and other elites' aggressive pursuit of diasporans, the quality of life improves, institutional capacities are strengthened, and infrastructures are enhanced. We have witnessed this progression from brain drain through ad hoc brain gain to brain circulation in nations that have advanced with the migration-development model and so did not simply leap from brain drain to brain circulation. The stay rate among nationals from the periphery who graduate from core universities was more than 90 percent where quality of life was poor (i.e., low life expectancy, high illiteracy, low GDP per capita, poor service delivery) and the atmosphere was politically repressive and culturally stultifying. It began to decline as more and more nationals began either to fully repatriate or to contribute in various ways with the encouragement of homeland political elites. At this stage, ad hoc brain gain was transforming into brain circulation. For a few years now, the stay rate among the most highly educated Koreans and Taiwanese in American universities has dropped below 50 percent because the quality of life in their respective homelands nears what they had experienced in the United States and because their homeland governments earnestly seek their counsel, capital, and other contributions.

A number of African nations are showing signs of transitioning out of brain drain to ad hoc brain gain because of improving economic conditions and growing political openness. Ghana and Botswana are chief among them. Again, African governments are requesting assistance from the IOM and institutions with similar missions, and this is encouraging. Also encouraging is the leadership on this issue shown by the African Union (AU). In 2003, the diaspora was declared the sixth region of the AU as a way to attract diasporan resources to the continent. Both the AU and NEPAD recognize that the economic, environmental, and social development many want for Africa cannot occur without active diasporan involvement. This is borne out by the fact that nations that repositioned themselves from being among the most impoverished and least served to being key players in the knowledge-intensive economy are actively involved with their respective diasporas in core countries.

Nevertheless, the diaspora has yet to be organically linked with AMCEN and NEPAD's Environment Initiative, either in the literature or in intergovernmental policy. The green migration-development model will allow leaders to focus their students and diasporas in core nations in policy areas that directly intersect with the economy concerning eco-industrial development. The fact is that technology and other forms of knowledge diffusion are as old as emigrating humanity itself (Diamond 1998). In the modern era, some European nations benefited from the migration-development model in the late nineteenth and early twentieth centuries, and some Asian nations, such as South Korea, Taiwan, India, and China, perfected the model in the late twentieth and early twenty-first centuries. Whereas the European experience is linked more to smokestack or basic industrial production, the Asian experience is inextricably linked with digital industrial production. Over the next couple of decades, Africa's economic transformation and its eco-industrial production could become linked—Africans could simultaneously produce, acquire, and transfer eco-industrial technology similarly to the way many Asian immigrants in the United States helped produce, acquire, and transfer digital technology. There is a need for AMCEN, NEPAD, the AU, and other African international institutions concerned equally with the environment, the diaspora, and socioeconomic development to give serious attention to this green migration-development strategy for eco-industrial development.

NEPAD, AMCEN, and African governments can encourage scientifically and technologically talented youths from across Africa to emigrate to the United States, Germany, Brazil, China, Denmark, Spain, and other nations presently leading in niched areas of eco-industrial R&D for study and work with the pioneers who will produce these next-era technologies. That a nation, region, or people engage in massive technology transfer to overcome a gargantuan technology gap in the migration-development model in no way means that its indigenous culture is deprivileged. In fact, if the indigenous knowledge systems of Africa are not organically integrated with state-of-the-art, cutting-edge, and next-era scientific and technological knowledge and tools, it is unlikely that the knowledge and tools will take hold. In a sense, then, productive indigenous knowledge systems are absolutely crucial. The idea is to jettison only the parts of these knowledge systems that locals see as no longer serving the public's interest and to adopt others that do.

We are witnessing other nations, erstwhile associates of the Global South, such as China, India, and Brazil, blazing new trails in the eco-industrial economy. China and India have their sights on leadership in renewable energy and are incentivizing their nationals with subsidies. China is already one of the largest national producers of solar photovoltaics (PVs) for the global market (Shi 2009). Zhengrong Shi is an example of successful brain circulation. He returned to China after earning a doctorate from Australia's University of New South Wales, studying at its Centre for Photovoltaic Engineering, bringing his knowledge and expertise with him and thus helping China become a world leader in solar power technology. In the process, he became one of the country's wealthiest citizens. Eco-industrial activities also result in jobs for Chinese citizens. Approximately eighty thousand people are employed in China's wind-turbine and solar-cell manufacturing industries. When the huge array of design, installation, and service provider jobs is considered, the number of Chinese eco-industrial employees increases to over six hundred thousand (Liu 2008).

South Africa also has industrial diasporans in the United States whose innovations that are contributing to the U.S. economy could also help enhance South Africa's economy. There is no better example than Elon Musk, founder and CEO of Tesla Motors—arguably the world's most celebrated electric car company—and SpaceX, a world-renowned rocket company that has received billions of dollars to launch rockets for NASA to the International Space Station. South Africa and other African nations have a lot to learn about wooing such innovative diasporans to contribute to their homeland development.

Bangladesh is a good example of eco-industrial progress because it illustrates that other nations of the Global South—beyond China, India, Brazil, and South Africa—can see immediate growth with renewable energy. "A project in Bangladesh, training local youth and women as certified solar technicians and as repair and maintenance specialists, aims to create some 100,000 jobs" (United Nations Environment Programme 2008).

African leaders, working in an enlightened framework and using concrete action plans, and with help from NEPAD, AMCEN, and other organizations, can prepare their citizens not only to protect and preserve Africa's ecosystems but also to be major contributors to the global green economy. Fortunately, many venture capitalists are looking at the green technology revolution as being even more lucrative for investors and of greater consequence for humanity than Internet-based firms. So the great race has begun for foreign green investment opportunities. The bigger point is that all of humanity will benefit from the contributions of Africans and all the world's peoples to the development of an eco-industrial economy.

The idea of Africa being a leader in some niched renewable energy sectors in the next decade or two is, for many, about as believable today as the idea in the mid-1960s of South Koreans becoming leaders in consumer electronics by the 1980s (Chang 2008). Certainly, it is possible for African nations to contribute to renewable energy technologies and be among their early adopters. And many Africans in the diaspora not only possess the human capital to invest in research, development, and commercialization of twenty-first-century eco-industrial technologies but also have accumulated significant economic and social capital from their years of working in high-tech institutions, universities, and R&D labs.

In the late 1990s and 2000s, I conducted systematic field studies of transnationalism in Singapore related to programs launched by the Singaporean government to assist its nationals studying in U.S. universities in STEM and business, particularly in Silicon Valley. A decade later I traveled to India to study its state-led efforts to advance through transnationalism, interviewing leaders in the Ministry of Overseas Indian Affairs (MOIA). Among the many objectives of MOIA is to encourage and support investment in India by affluent Indian nationals abroad. Insights I gleaned from those interviews were helpful in subsequent field trips between 2009 and 2014 to research the African diaspora and renewable energy in Southern Africa. In October 2009, I attended a major renewable energy conference in Johannesburg, co-organized by the International Solar Energy Society and the Solar Energy Society of Southern Africa. Renewable-energy technologists, government environmental affairs officials, sustainability entrepreneurs, and others from more than eighty countries participated, and I interviewed many of them on best practices, including engaging diasporans living and working in countries at the forefront of renewable-energy R&D. I also discussed prospects for strategic implementation of a green migration development model with African leaders while attending the 2011 United Nations Framework Convention on Climate Change in Durban, South Africa. I learned that leaders seem supportive of the idea but that funding it is problematic. Africana studies can contribute to this project by further developing the literature of African transnationalism and by integrating the literatures of Africana and environmental studies.

Renewable Energy in Africa

Renewable energy currently powers only 13 percent of the world's economy. However, rightful fears of climate change, peak oil production, fracking, and general fossil fuel burning, combined with a perhaps overexuberant anticipation of jobs in renewable energy industries and other parts of the burgeoning green sector, suggest that renewable energy as a percentage of global energy use will continue to surge at the expense of fossil fuels. This could be hugely beneficial for Africans ecologically, biologically, and economically, though it will hurt African oil-exporting states such as Nigeria, Angola, Equatorial Guinea, and South Sudan. As the world transitions to renewable energy platforms, the question is whether Africans will have the imagination, the will, and the strategies to lead.

Renewable energy sources/technologies include solar thermal, PV, biofuels, wind power, geothermal, tidal power, and wave power. These can be divided into old and new. Old renewable energy sources—biomass and hydropower—have been around for quite some time and do not necessarily depend on advanced science and technology. Because of their reliance on these sources, the share of renewable energy in non-OECD (Organisation for Economic Co-operation and Development) countries is almost four times higher than that in OECD countries.

Only about a third of sub-Saharan Africans have access to electric grids. Obviously without electricity there is no way to store vaccines, run computers, or study at night, all of which cripples a country's efforts to become a major participant in the global economy. Another downside of marginal electrification, particularly in rural areas, is poor health caused by reliance on biomass and charcoal cooking. Table 5.2 shows the high percentage of energy supplied by biomass burning in African nations. Every year, millions die from respiratory ailments from accumulated smoke inhalation associated with biomass-fueled cooking. Such an important fact should give pause to individuals who, for some inexplicable reason, are not particularly keen on renewable energy.

Africa's rates of electrification resemble its rates of fixed telephone line connectivity (also shown in Table 5.2) in the sense that both are minuscule in comparison to those in more developed nations. Just as

TABLE 5.2 AFRICAN ELECTRIFICATION RATES, RENEWABLE ENERGY USE, AND TELEPHONE ACCESS (SELECTED COUNTIES)

Country	Electrification rates, 2000–2005 (%)	Hydro, solar, wind, geothermal energy, 2005 (%)	Biomass and waste, 2005 (%)	Fixed telephone lines per 100 inhabitants, 2008	Mobile telephone subscriptions per 100 inhabitants, 2008
Algeria	98	0.1	0.2	9.6	92.7
Angola	15	1.5	63.8	0.6	37.6
Benin	22	—	64.7	1.8	39.7
Botswana	39	—	24.1	7.4	77.3
Burkina Faso	7	—	—	1.0	16.8
Cameroon	47	4.8	78.6	1.0	32.3
Congo	20	2.5	58.3	0.6	50.0
Congo (DRC)	6	3.7	92.5	0.1	14.4
Cote d'Ivoire	50	1.6	58.3	1.7	50.7
Egypt	98	1.9	2.3	14.6	50.6
Eritrea	20	—	64.8	0.8	2.2
Ethiopia	15	1.1	90.6	1.1	2.4
Gabon	98	4.1	58.8	1.8	89.8
Ghana	49	5.1	66.0	0.6	49.6
Kenya	14	5.9	74.6	0.6	42.1
Lesotho	11	—	—	3.2	28.4
Madagascar	15	—	—	0.9	25.3
Malawi	7	—	—	1.2	12.0
Mauritius	94	—	—	28.5	80.7
Morocco	85	1.0	3.3	9.5	72.2
Mozambique	6	11.2	85.4	0.4	19.7
Namibia	34	10.3	13.5	6.6	49.4

(continued)

TABLE 5.2 *(continued)*

Country	Electrification rates, 2000–2005 (%)	Hydro, solar, wind, geothermal energy, 2005 (%)	Biomass and waste, 2005 (%)	Fixed telephone lines per 100 inhabitants, 2008	Mobile telephone subscriptions per 100 inhabitants, 2008
Nigeria	46	0.7	78.0	0.9	41.7
Senegal	33	2.0	39.2	2.0	44.1
South Africa	70	0.2	10.5	8.9	90.6
Sudan	30	0.6	79.5	0.9	29.0
Tanzania	11	0.7	92.1	0.3	30.6
Uganda	9	—	—	0.5	27.0
Zambia	19	10.7	78.7	0.7	28.0
Zimbabwe	34	5.2	61.9	2.8	13.3

Note: Data from United Nations Development Programme, *Human Development Report*, 2000–2006; International Telecommunication Union, *World Telecommunication/ICT Indicators Database*, September 2008. Dashes indicate where data are unavailable.

the cost of massive landline infrastructure rollout alterations proved prohibitively expensive in most African countries, so too might be the building of massive electrification grid infrastructure, particularly as Africa has a high percentage of rural populations. Just as Africans are gaining telecom access by bypassing landlines and leapfrogging directly into mobile telephony, so are vast numbers of them likely to gain electrification by bypassing centralized fossil fuel–based grids and leapfrogging directly to decentralized renewable energy sources.

In the mid-1990s, there were more phones in Manhattan than on the entire African continent. Two decades later, because of the telecom technological leapfrogging, Africa now has more subscribers than America or Europe. The only question is whether Africans will be among the pioneers of renewable energy technological leapfrogging or very early or very late adopters. Some renewable energy technologies, such as solar cookers and biogas, already deliver power to rural Africans and thus allow them to complete simple yet crucial tasks without harm to their lungs and eyesight. Solar cookers are often just parabolic mirrors that concentrate solar thermal energy at a focal point. With them, water can be boiled in minutes, and rice and other foods can be cooked in about the same time it takes with natural gas or electric stoves. Most important, they do not emit carbon and are not harmful because they create no smoke.

Solar cookers can be purchased for less than $300 and can last more than ten years. They are an "appropriate technology" product that is somewhat simple to build and requires minimal training to operate. They also contribute jobs in production, distribution, training, and repair. In Bangladesh, for example, 100,000 new jobs are anticipated for youths and women as certified solar technicians and repair specialists. India plans to create 150,000 jobs by replacing inefficient biomass cooking stoves in 9 million households. If India were to scale up this effort to meet the vast energy needs of its entire population, there might be millions of new jobs. Ultimately, for an investment of about $30 a year for each family, Africa can improve the environment, public health, local economies, and jobs prospects.

Biogas, another decentralized energy source, is ideal for rural Africa. Biogas technology converts cattle dung, chicken droppings, and human body waste into electricity. A biogas digester system may cost approximately $700. An important study in Uganda determined that the following factors probably caused changes in households adopting biogas technology: decreased age of head of household, increased household income, increased number of cattle owned, increased household size, increase in male heads of household, and increased cost of traditional fuels (Walekhwa, Mugisha, and Drake 2009). Africans' contributions to the development of this evolving technology have been minimal.

Although solar cookers and biogas electricity are more affordable, PV technology is also a major potential supplier of energy for rural Africa. Many African nations have promoted solar PV, particularly for rural households, but few have had much success. Cost has been a factor, but so too has the lack of sufficiently trained maintenance and repair specialists. A kilowatt system can cost several times a household income. Ironically, plummeting PV prices may, at least initially, increase global energy inequality in that, as core and peripheral countries add PV technology to their energy portfolios, prices drop to thresholds of affordability and costs rivaling heavily subsidized fossil fuels. These countries could begin incorporating more PV technology into their energy portfolios while many African countries could not, since their affordability levels are significantly below those of the core and semi-peripheral nations, thereby initially widening the gap. As other nations begin to rely more on PV energy, both on-grid and off-grid, some African nations will still be waiting for PV prices to drop. We witnessed this phenomenon with mobile telephones and other information and communication technologies (ICTs) for developing countries. Fortunately, once prices began tumbling in the core, they fell somewhat quickly to affordable levels in developing regions, including Africa (Patterson and Wilson 2000). Nations of the Global South that had the leadership, imagination, and strategies for migrating from a sparse landline network to a national mobile platform became early adopters as well as some of the new technology's greatest beneficiaries. This potential affordability problem underscores the necessity for Africans to position themselves through the migration-development model and sustainability entrepreneurship to help drive renewable energy innovation.

Renewable energy is presently a minor part of the total global energy picture, but it is on track to become the principal source of energy to power the global economy. Figure 5.1 shows the low contribution of renewable energy—only 9.1 percent—relative to fossil fuels in the United States in 2011 (U.S. Energy Information Administration 2012a). Globally, among renewable sources, biomass was 48 percent, and hydroelectric was 35 percent. This ratio will qualitatively change, and renewable energy will become the dominant energy source, but it will likely be a long-term proposition, and the timing of this transformation cannot be accurately predicted. What we can strongly assert is that when the energy inflection point is reached, it will largely be a function of public and private investment in renewables research, development, commercialization, and deployment. Governments in the Global North as well as in China and Brazil are ramping up their renewables R&D funding. In 2011, the United States reclaimed from China status as the country with the largest investment in clean

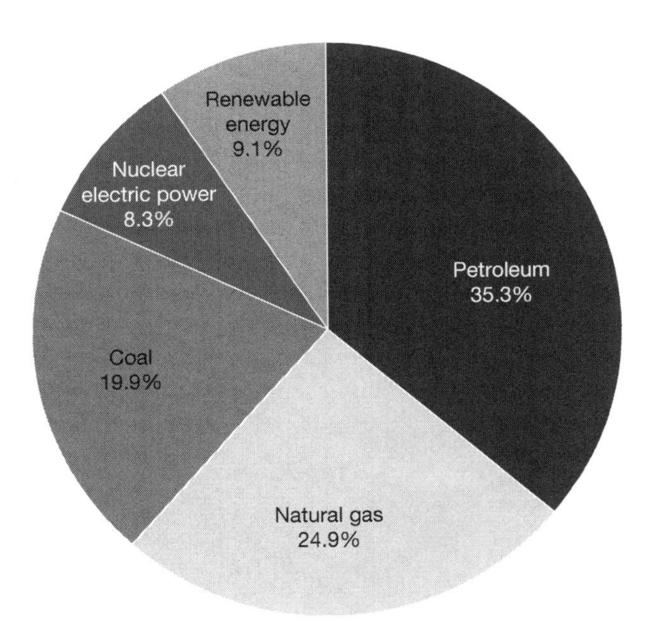

Figure 5.1 Primary U.S. energy consumption in 2011 by source. *(U.S. Energy Information Administration 2012b)*

energy, with spending totaling $55.9 billion (Bloomberg New Energy Finance 2012), although that was a result of the massive one-off stimulus package in the wake of the financial crisis. Additionally, renewable energy is the fastest-growing area for global venture capitalist commercialization. If Africans hope to claim a renewable energy niche, they will have to jump into the middle of this ferment of research, development, and commercialization.

Succeeding with Eco-Development in Africa Will Be Challenging

It is best to conclude this chapter with more sobriety on the prospects of eco-industrial development in Africa. I am enthusiastic about Africa's potential successful use of a green migration development model, in which African governments and diasporas work together strategically in the interest of the homeland, but this is only possible, not necessarily probable. While it is true that African governments are held accountable more so now than at any time since independence, it is often difficult to see how they will be able to form and execute policy effectively and efficiently to meet the needs of their people at the level required for the green migration development model to work. A senior executive at Nedbank, one of South Africa's largest banks, told me in an interview in October 2013 in a Johannesburg suburb that "Africa's economies are not so much emerging economies as they are frontier economies." "Frontier" in this context represents the border between order and chaos—that is, between rational predictability and unpredictability. Although African countries are on the whole better run today and have much more potential than they have had for many decades, it seems unwise to assert that the impressive progress over the past fifteen years or so will absolutely continue on its upward trajectory. For that matter, it is not a given that even China will absolutely continue on its stunningly successful trajectory that began merely two decades ago.

History is rife with contingencies. One of them is corruption, which has to be more effectively contained. Some progress has been made, but much more is needed. Corruption is the abuse of public office not only for personal gain but also through public policies that benefit the elites while in effect ignoring the poor. This form of corruption is probably the greatest challenge to economic progress and sustainable economic transformation. I hasten to add that all nations are corrupt in that their governments are to varying degrees increasingly being captured by elite interests. By this reasoning, all nations should be regarded as developing because "development is about changing a society to enhance people's well-being across generations—enlarging their choices in health, education and income and expanding their freedoms and opportunities for meaningful participation in society" (United Nations Development Programme 2013: 66).

Since no nation has categorically achieved this, all nations are developing in various degrees. That said, some nations can absorb corruption more than others. Take the United States. It has a $17 trillion economy, and when, say, $20 billion is lost because of corruption, the loss is more easily absorbed by and almost imperceptible to American society. In addition, the country already has world-class infrastructure—although it is due for a twenty-first-century upgrade—in transportation, communication, health care, education, governance, research and development, and so on. Comparing losses to corruption in a country such as the United States with losses to corruption in Africa makes for a stark contrast. Even $500 million, $100 million, or $50 million lost to corruption can be crippling to a developing economy and broader society, given that a quarter of the population might be illiterate, more than 40 percent might be unconnected to an electric grid or be without modern power, and the government might be spending almost no money on research and development. For African nations such a loss would be disastrous because those funds could have gone to build the basic infrastructure and other necessities that a country like the United States takes for granted.

African nations have so much ground to make up to become effective competitors in the global marketplace. Even if they had no corruption and did everything "correctly," there would still not be enough resources to improve the quality of life for all of their citizens in the intermediate period in a number of countries. Because, like so many other countries, African nations do not have resources or time to waste, they need a developmental state apparatus in which technocrats who put national development above personal gain can advance agendas such as a green migration-development model. By definition, "a state is developmental when it establishes as its principle of legitimacy its ability to promote and sustain development, understanding by development the combination of steady high rates of economic growth and structural change in the productive system, both domestically and in relationship to the international economy" (Castells 1992: 56). The following features empirically distinguish developmental states from other Global South nation-states: (1) state-business reciprocity, (2) high-performance bureaucracy, (3) corruption containment, and (4) entrepreneurial technology promotion (Robinson and White 1998; Leftwich 1995).

Developmental states, as originally conceptualized by Chalmers Johnson (1982), privilege economic development for state support to the exclusion of other factors. Their governments identify strategic industrial priorities and then mobilize national resources and public, private, and nonprofit resources to meet them. Developmental states are moreover known for high-performance bureaucracies that efficiently channel resources toward prespecified priorities to generate targeted outcomes. Again, corruption exists in all societies, including developmental states, but the latter tend to hand down swifter and harsher punishment. More important, corruption does not prevent the state from fulfilling its end of the social contract—ensuring improvement of the lives of its citizens. One means of minimizing corruption is to pay bureaucrats more than they might command in the private sector, with the understanding that they will be cashiered as swiftly as, if not more swiftly than, they would in the private sector for inadequate performance. If African nations were to become developmental states, their prospects for advancement in eco-technological development would be substantially increased.

Although African nations have entertained notions of becoming developmental states, none have succeeded. By definition, "in [Africa], a development[al] state needs to be capable of planning and managing investment in sectors normally neglected by private investors but essential for 'a higher skill, quality based export trajectory.' In a developmental state, political leaders and bureaucrats must be capable of resisting sectional pressure" (Lodge 2009: 253). Political leaders such as former South African president Thabo Mbeki, have talked about the need for African developmental states, but none exist, and scholarship on this issue has been thin (Patterson 2001). Some see the reason as "timing is everything," postulating that the opportunity to create a developmental state was terminated with the arrival of the transnational capitalist class in the late 1980s and its push for globalization from then onward. The emergence of transnationalism was an important inflection point in history. Any nation that had not become developmental by that time was unlikely to do so in the future.

Three important observations need underscoring. First, African nations on the whole are undergoing a transformational change in terms of democratic and bureaucratic accountability, as well as smarter economic policies resulting in faster, although sadly not inclusive or widespread, economic growth. Second, Africans leapfrogged over vintage technologies like landline telephony, so they can also leapfrog over economic bases to reach cutting-edge eco-industrial production. Such an outcome is more plausible for some African nations than for others. An African developmental state would likely be the best political and economic means of achieving it. Third, African developmental states arguably have not existed. And now that we are in the era of a transnational capitalism and globalization, it is unlikely that a new developmental state apparatus will develop for African or other nations. Still, if African nations could incorporate some of the attributes of a developmental state, such as corruption containment and a high-performance bureaucracy, they would be better positioned to produce leaders in sustainable economic development. At base, Africa needs its diaspora to help it build green economies in sustainable societies, and a green Africana studies would be best equipped to eventually position the diaspora to do so.

VIGNETTE 5.1. CROSBY MENZIES
Solar Energy in Africa

Solar energy offers Africa its best opportunity for lifting itself out of poverty while simultaneously leapfrogging dirty and costly mistakes made by industrialized and polluted nations. Globally, fossil fuels are still being "justified" as a necessity to power growth and development, despite much better and completely clean alternatives that have been available for over eighty years.

Africa remains the least electrified continent on Earth, making it the *ideal place to introduce state-of-the-art solar technologies. It has been proven that solar energy can easily provide for all humanity's energy needs with thousands of times more energy arriving on the planet than we collectively use in a year, free of charge!*

The continent urgently needs to create enabling policies and to develop local and international science and technology centers dedicated to this issue. Africa also needs to divert a considerable amount of investment into renewables, an area in which it is sadly lagging behind. Meanwhile, imports and subsidies of fossil fuels represent a major drain on African resources.

SunFire Solutions was founded in 2004 to promote, manufacture, and distribute solar cookers and solar lighting, and to spread the word about renewable energies in general. Africa's sparsely populated landscape lends it to decentralized or no-grid solutions, which frustrated the old energy platform. No-grid solutions make it possible to provide energy to all households.

The continent is literally drenched in sunlight, with even the rainy equator belts offering at least four hours of sunlight each day, which is ample power for lights, mobile phones, and solar hot-water systems. SunFire decided to take this quiet crisis and turn it into an opportunity by opening a company in 2004 to focus on addressing the energy challenge, which is one of the greatest on the continent. At present, firewood for cooking is currently the single biggest energy source in Africa. Biomass burning harms respiratory systems, eyesight, and other areas of health for millions of Africans. Solar cookers, when combined with fuel-efficient wood-burning stoves and retained heat bags, offer users an 80 percent reduction in cooking fuel from day one. Deforestation represents one of the biggest threats to our continent's biodiversity, which draws many millions of tourist dollars and is set to grow if preserved.

Many of the communities where SunFire works are extremely impoverished, meaning that introducing technologies via a company (i.e., buying and selling) is not an option. To reach these communities,

SunFire has registered a sister NGO to provide solar cookers at low or no cost.

Cooking goes right to the heart of many households, and it is SunFire's core strategy to address the immediate energy needs of African communities using renewable energies while creating thousands of job opportunities.

Two questions that must be asked are why there is so little solar technology available to Africa's one billion citizens and why there is such minuscule support for solar technology from Africa's governments, when sometimes as much as 25 percent of a country's GDP leaves the continent to import fossil fuels.

We don't have answers to this, but we know what the solution is: by using the sun, Africa can rightfully take its place as the continent of light and enjoy a bright future.

Conclusion

In writing this book, I set out to fulfill straightforward objectives and complete a challenging task. One of my objectives was to provide evidence that unquestionably establishes Africana studies as a rigorous and hugely important academic field but also to show that it has not yet invested nearly enough of its scholarly attention in environmental degradation and climate change, which adversely affect persons of African heritage more than they do other groups. The empirical data presented in Chapter 1 support this proposition. For example, I looked at self-reported data on Africana studies programs and their faculty all across the country to determine if the environment was part of their central focus. I also examined course inventories to glean whether the programs were exposing students to systematic knowledge about the environment in general as it relates to the black community in particular. Further, I examined the articles in the field's top three journals (which Africana studies scholars regularly read and direct their students to read) over a fifteen-year period to establish the frequency of peer-reviewed research on environmental issues. Seeking evidence to refute the proposition that Africana studies has given insufficient attention to the environment, I even tried contacting major publishers of books on the environment to discover whether Africana studies professors were adopting them for their courses. Apparently publishers do not divulge such information, presumably for proprietary and privacy reasons. Whatever the reasons, such information could further corroborate my proposition or perhaps militate against it.

In Chapter 1, I also offer a few important caveats regarding my belief that Africana studies should have more to do with environmental issues. For instance, I acknowledge that leading scholars on the environment and people of color, such as Dorceta Taylor and Robert Bullard, have had their works published in journals other than the three major Africana studies publications and by book publishers other than the African World Press and Third World Press. Furthermore, these scholars may not have presented at Africana studies conferences any of their considerable research on the intersection of race and the environment, particularly at those sponsored by the National Council for Black Studies and the Association for the Study of African American Life and History. Nevertheless, their works incorporate the sensibilities of the general framework of Africana studies. That is, their scholarly mission is in keeping with the mission of Africana programs. The academic gap between Africana and environmental programs therefore remains a lost opportunity for students because they are not introduced to the serious environmental scholarship of Taylor and Bullard and others.

Another caveat is that some Africana studies programs and professors may indeed cover the environment, but there may be no indication of such coverage in program descriptions or course titles, once again suggesting the insufficient attention to the environment in Africana studies. In fact, today's environmental challenges and opportunities are so great that they warrant dedicated courses and centrality in Africana studies. The program I directed at the University of Toledo was among the first, if not the first to give environmental challenges and opportunities in the black community—primarily in the United States and Africa—a central position. Locally, we mostly engage with community-based organizations in brownfield remediation, urban agriculture, off-grid solar photovoltaics in Southern Africa, and environmental awareness efforts. Globally, we engaged in renewable energy policy and sustainability and social entrepreneurship in Southern Africa. The hope is that a great number of Africana studies programs and scholars will soon cover the environment for its relevance to the black community as extensively and as brilliantly as they now cover history, culture, and the arts.

A straightforward objective in completing this book was to demonstrate the huge and grinding need in black communities to address their environmental challenges. Another was to emphasize the need for more scholars, activists, and other serious professionals with the skills and inclination to help the community with its environmental problems. Chapter 3 examines a few of the major types of locally unwanted land uses, or LULUs—namely, brownfields, toxics release inventory (TRI) facilities, and Superfund (hazardous waste) sites. I drill down into the substantial and well-documented needs of the black community with respect to LULUs. Africana studies and many of its scholars have an abiding interest in addressing the historical challenges faced by the global black experience and in addressing new ones. From the classical African societies that European and other invaders demolished through enslavement and colonization to the black power movement and to the horrific failure of schools to meet the needs of black children, Africana scholars have been on a mission to better document, reclaim, and draw meaning from the black experience and to empower black people to pursue redemption, redress wrongs, and work to build a more equitable world. Discussions of environmental problems and their responsibility for stunting the prospects of black people, even in utero and throughout their prematurely cut-short lives, are worthy of the same intellectual and visceral need for engagement and are covered in Chapter 3.

As I emphasize throughout, studying the environment is not only about addressing monumental problems but also about pursuing monumental opportunities. Green jobs exist and are only going to explode over the long haul to help us avoid an environmental catastrophe. These gains may come in two forms: (1) mitigating and adapting to the environmental problems wrought by brown industrialization, which focuses on labor productivity rather than resource productivity, and (2) innovating green futures for an economy that contains little or no waste, carbon emissions, or other toxins. Some green jobs will be blue collar, and others white collar. Many will require little advanced skill—for example, driving waste and carbon-free public transportation vehicles; many will require huge investments in education and training—for example, developing energy storage technologies for intermittent renewable sources such as solar and wind. Most green jobs will be situated between these two extremes: they will require more than a secondary education but not necessarily a four-year degree.

Environmental justice scholars have challenged federal, state, and local agencies, as well as mainstream environmental NGOs, about the paucity of blacks and other people of color they employ, particularly in senior decision-making positions. These agencies and NGOs transitioned from their initial defensive posture of asserting that they are not racists to asserting that blacks primarily but also others of color have little or no interest in the environment to asserting that such individuals lack the training in fields such as environmental policy, sciences, law, engineering, and natural resource management. There is some merit in the claim of insufficient training. It is rightly looked on as a problem in that many blacks and others of color disproportionately lack the skills and credentials to be hired for green professional positions.

While green Africana studies is critical to sparking African Americans' deeper interest in environmental issues and careers, opportunities for Africana students to obtain the education they need to succeed in these careers are limited, since Africana programs exist at predominantly white institutions where some 85 percent of African American students matriculate. Although black students do enroll in Africana studies courses at those institutions, they may rarely take environmental science courses. Africana courses that are dedicated to environmental matters and others that incorporate environmental concerns will likely stimulate students to enroll in related courses such as natural resource management, environmental sciences, and urban ecology. Graduates who have taken such courses will not only get some of those green jobs but also help create them because they are better able to become sustainability entrepreneurs and public policy administrators in green agencies at various levels of government.

Although much of the book has an African American focus, Chapter 5 is dedicated to environmental challenges and career opportunities in Africa. In Africa's colonial and postcolonial experience, only a minority of continental Africans received benefits from fossil fuel–based industrialization, and many suffered severely, and disproportionately, from the environmental degradation wrought by that industrialization. Now, for the first time, Africans are systematically and materially benefiting from economic gains: child mortality on the continent is dropping at a record pace, the middle class is expanding and deepening at unprecedented levels, and the economy is shifting from one in which citizens eke out an existence to one in which citizens produce in quantities to satisfy larger shares of the public market, thereby giving people choices in their lives. A major downside of Africa's economic transformation is that the environmental problems triggered by the broad use of environmentally degrading and climate-altering technologies, in far too many instances, continue to wreak havoc on African societies. As in other parts of the world, droughts and floods are now occurring more frequently and more devastatingly in various regions of the continent. Africans may be responsible for only about 2 percent of global carbon emissions, but they suffer the most from climate change, primarily because of higher levels of poverty and shortages of resources ranging from technical expertise to basic technology.

The good news out of Africa nowadays is not just about plummeting child mortality rates and the transforming economy. It is also about strengthening institutions and expanding imagination when it comes to environmental challenges and opportunities. The African Ministerial Conference on the Environment (AMCEN), the African Union (AU), and the New Partnership for Africa's Development (NEPAD) are among high-level African institutions now focusing on mitigation of and adaptation to environmental problems. This is because resources are in short supply to deal with these massive problems, which express themselves regionally, not just nationally. For these reasons, AMCEN and NEPAD foster collaboration to avoid costly blind spots and unaffordable duplication. As I discuss in Chapter 5, collaborative approaches are absolutely essential, but they are more defensive than offensive.

A more forward-leaning development strategy is what I call the green migration-development model, or green African transnationalism. In other words, Africans are learning from the playbooks of Asian and other nations that have successfully engaged their diaspora in transnational projects to support the homeland. East Asians, to take one example, centered their models on information and communications technologies, whereas Africans might center theirs on green technologies. African transnational development involves the strategic migration of African nationals to nations that are pioneering green technology, including renewable energy, energy efficiency, and zero-waste technologies. It is also engaging generations-long diasporans who have the expertise as well as the inclination to be supportive of Africa's green transformative development.

In the opening sentence of this Conclusion, I state that I had some straightforward objectives to fulfill, which have been summarized in the preceding paragraphs. I also took on a challenging task: linking disparate academic fields—Africana studies and environmental studies—that up to now have had little interaction. As discussed mainly in Chapter 2, the question is whether the intellectual concerns of environmental studies, including those of scholars such as Bullard and Taylor, can comport with the agenda and knowledge frame of Africana studies. There had never been, before this study, a book-length argument for a formal intellectual marriage between Africana studies and environmental studies. Bridging literatures can be tricky and often meets with failure, but when they are linked organically and successfully, both literatures and the broader public tend to benefit. In Chapter 2, I make an opening attempt at joining these two important disciplines through my extended analysis of their common paradigmatic approaches.

Over time, the literature on the greening of Africana studies will undoubtedly grow in size, scope, and sophistication. Black communities in cities where Africana studies programs are a resource for addressing local environmental concerns will benefit. American society will benefit as more African Americans and others become passionate about the threats to the environment and gain decision-making positions that allow them to act on their social equity sensibilities. And Africa will benefit from the greater number of trained experts eager to participate in green African transnational development projects. With the organic integration of Africana studies and environmental studies, better and indeed greener days are ahead in the black community and the global society.

References

Adelaja, Soji, Judy Shaw, Wayne Beyea, and J. D. Charles McKeown. 2010. "Renewable Energy Potential on Brownfield Sites: A Case Study of Michigan." *Energy Policy* 38:7021–7030.

Advanced Research Projects Agency–Energy. 2012. "Director Majumdar Testifies on Fiscal Year 2013 Budget Request." March 28. Available at http://arpa-e.energy.gov/?q=arpa-e-news-item/director-majumdar-testifies-fiscal-year-2013-budget-request.

African Ministerial Conference on the Environment. 2006. *History of the African Ministerial Conference on the Environment: 1985–2005*. Nairobi, Kenya: AMCEN.

Akinyela, Makungu M. 1995. "Rethinking Afrocentricity: The Foundation of a Theory of Critical Africentricity." In *Culture and Difference: Critical Perspectives on the Bicultural Experience in the United States*, edited by Antonia Darder, 21–40. Westport, CT: Bergin and Garvey.

Alkalimat, Abdul. 2001. "Toward a Paradigm of Unity in Black Studies." In *The African American Studies Reader*, edited by Nathaniel Norment, 480–495. Durham, NC: Carolina Academic Press.

Alkalimat, Abdul, and Kate Williams. 2001. "Social Capital and Cyberpower in the African-American Community: A Case Study of a Community Technology Centre in the Dual City." In *Community Informatics: Shaping Computer-Mediated Social Relations*, edited by Leigh Keeble and Brian D. Loader, 177–204. New York: Routledge.

Asante, Molefi Kete. 1996. "The Principal Issues in Afrocentric Inquiry." In *African Intellectual Heritage: A Book of Sources*, edited by Molefi Kete Asante and Abu S. Abarry, 256–261. Philadelphia: Temple University Press.

Association for the Advancement of Sustainablity in Higher Education. 2009. "Annual Report, 2009." Available at http://www.aashe.org/files/AASHE_AnnualReport(3).pdf.

Atlantic Council. 2010. "Developing a Realistic and Balanced United States Electric Power Generation Portfolio: Assuring Energy, National, Economic and Environmental Security." Available at http://www.atlanticcouncil.org/images/files/publication_pdfs/403/AtlanticCouncil_USElectricPowerGenerationPortfolio.PDF.

Ayittey, George B. N. 2006. *Africa Unchained: The Blueprint for Africa's Future*. New York: Palgrave Macmillan.

Bacot, Hunter, and Cindy O'Dell. 2006. "Establishing Indicators to Evaluate Brownfield Redevelopment." *Economic Development Quarterly* 20 (2): 142–161.

Bae, Hyunhoe. 2012. "Reducing Environmental Risks by Information Disclosure: Evidence in Residential Lead Paint Disclosure Rule." *Journal of Policy Analysis of Management* 31 (2): 404–431.

Bambara, Toni Cade. 1970. *The Black Woman: An Anthology*. New York: Penguin.

Barringer, Felicity. 2012. "For New Generation of Power Plants, a New Emission Rule from the E.P.A." *New York Times*, March 27. Available at http://www.nytimes.com/2012/03/28/science/earth/epa-sets-greenhouse-emission-limits-on-new-power-plants.html?_r=1&.

Beder, Sharon. 1998. *Global Spin: The Corporate Assault on Environmentalism*. White River Junction, VT: Chelsea Green.

Bennett, Lerone. 1984. *Before the Mayflower: A History of Black America*. New York: Penguin.

"The Best Story in Development." 2012. *The Economist*, May 19, p. 56.

Biondi, Martha, ed. 2012. *The Black Revolution on Campus*. Berkeley: University of California Press.

Birol, Fatih. 2006. "World Energy Prospects and Challenges." *Australian Economic Review* 39 (2): 190–195.

"The Black Radical Tradition, Progress and Marxism." n.d. *Global South* blog. Available at http://globalsouth12.wordpress.com/the-black-radical-tradition-progress-and-marxism/ (accessed October 10, 2012).

Blauner, Robert. 1972. *Racial Oppression in America*. New York: Harper and Row.

Bloomberg New Energy Finance. 2012. "Solar Energy Surge Drives Record Clean Energy Investment in 2011." January 12. Available at http://www.bnef.com/PressReleases/view/180.

Boggs, Grace Lee. 2011. *The Next American Revolution: Sustainable Activism for the Twenty-First Century*. Berkeley: University of California Press.

Brender, Jean, Julian A. Maantay, and Jayajit Chakraborty. 2011. "Residential Proximity to Environmental Hazards and Adverse Health Outcomes." *American Journal of Public Health* 101 (S1): S37–S52.

Brewer, Rose M. 2003. "Black Radical Theory and Practice: Gender, Race, and Class." *Socialism and Democracy* 17:109–122. Available at http://www4.uwm.edu/c21/conferences/2008since1968/brewer_blackradicaltheory.pdf.

Brynjolfsson, Erik, and Andrew McAfee. 2012. *Race against the Time Machine: How the Digital Revolution is Accelerating Innovation, Driving Productivity, and Irreversibly Transforming Employment and the Economy.* Lexington, MA: Digital Frontier Press.

———. 2014. *The Second Machine Age: Work, Progress, and Prosperity in a Time of Brilliant Technologies.* New York: W. W. Norton.

Buffett, Warren. 2003. "Berkshire Hathaway Inc. 2002 Annual Report." Available at http://www.berkshirehathaway.com/2002ar/2002ar.pdf.

Bullard, Robert D. 1990. *Dumping in Dixie: Race, Class, and Environmental Quality.* Boulder, CO: Westview Press.

———. 1993. "Anatomy of Environmental Racism." In *Toxic Struggles: The Theory and Practice of Environmental Justice,* edited by R. Hofrichter, 25–35. Philadelphia: New Society.

———. 2005. "More Blacks Overburdened with Dangerous Pollution: AP Study of EPA Risk Sources Confirms Two Decades of EJ Findings." Environmental Justice Resource Center, December 19. Available at http://www.ejrc.cau.edu/BullardAPEJ.html.

———, ed. 2007. *Growing Smarter: Achievable Livable Communities, Environmental Justice, and Regional Equity.* Cambridge, MA: MIT Press.

Bullard, Robert D., Paul Mohai, Robin Saha, and Beverly Wright. 2007. *Toxic Wastes and Race at Twenty: 1987–2007.* Cleveland, OH: United Church of Christ Justice and Witness Ministries. Available at http://www.ucc.org/justice/pdfs/toxic20.pdf.

Bureau of Labor Statistics. 2012. "Employment in Green Goods and Services—2010." Available at http://www.bls.gov/news.release/archives/ggqcew_03222012.pdf.

Burger, Joanna, and Michael Gochfeld. 2011. "Conceptual Environmental Justice Model for Evaluating Chemical Pathways of Exposure in Low-Income, Minority, Native American, and Other Unique Exposure Populations." *American Journal of Public Health* 101:S64–S73.

Cabral, Amilcar. 1973. *Return to the Source: Selected Speeches of Amilcar Cabral.* New York: Monthly Review Press.

Calderón-Garcidueñas, L., R. Engle, A. Mora-Tiscareño, M. Styner, G. Gómez-Garza, H. Zhu, V. Jewells, et al. 2011. "Exposure to Severe Urban Air Pollution Influences Cognitive Outcomes, Brain Volume and Systemic Inflammation in Clinically Healthy Children." *Brain Cognition* 77:345–355.

Callicott, J. Baird. 1987. *Companion to a Sand Country Almanac.* Madison: University of Wisconsin Press.

Carmichael, Stokely, and Charles V. Hamilton. 1967. *Black Power: The Politics of Liberation in America*. New York: Vintage.

Carson, Rachel. 1964. *Silent Spring*. Greenwich, CT: Fawcett.

Castells, Manuel. 1992. "Four Asian Tigers with a Dragon Head: A Comparative Analysis of the State, Economy, and Society." In *Development in the Asian Pacific Rim*, edited by Richard Applebaum and Jeffrey Henderson, 33–70. Newbury Park, CA: Sage.

Center for Responsible Travel. n.d. "Responsible Travel: Global Trends and Statistics." Available at http://www.responsibletravel.org/news/Fact_sheets/Fact_Sheet_-_Global_Ecotourism.pdf (accessed August 1, 2014).

Chang, Ha-Joon. 2008. *Bad Samaritans: The Myth of Free Trade and the Secret History of Capitalism*. New York: Bloomsbury Press.

Chapple, Karen, Cynthia Kroll, T. William Lester, and Sergio Montero. 2011. "Innovation in the Green Economy: An Extension of the Regional Innovation System Model?" *Economic Development Quarterly* 25:5–25.

Chavis, Ben, ed. 1987. *Toxic Wastes and Race in the United States: A National Report of the Racial and Socio-economic Characteristics of Communities with Hazardous Sites*. Cleveland, OH: United Church of Christ Commission for Racial Justice.

Chilonda, Pius, Charles Machethe, and Isaac Minde. 2007. "Poverty, Food Security and Agricultural Trends in Southern Africa." ReSAKSS Working Paper No.1, Regional Strategic Analysis and Knowledge Support System for Southern Africa, Pretoria, South Africa. Available at http://pdf.usaid.gov/pdf_docs/PNADS604.pdf.

CIA. 1995–1996. *World Factbook*. Washington, DC: Central Intelligence Agency.

———. 1997. *World Factbook*. Washington, DC: Central Intelligence Agency.

———. 2007. *World Factbook*. Washington, DC: Central Intelligence Agency.

Collard, Andree. 1988. *Rape of the Wild: Man's Violence against Animals and the Earth*. Bloomington: Indiana University Press.

Collins, Flannary P. 2003. "The Small Business Liability Relief and Brownfields Revitalization Act: A Critique." *Duke Environmental Law and Policy Forum* 13 (2): 303–328. Available at http://scholarship.law.duke.edu/cgi/viewcontent.cgi?article=1132&context=delpf.

Collins, Patricia Hill. 1991. *Black Feminist Thought: Knowledge, Consciousness, and the Politics of Empowerment*. New York: Routledge.

Cooper, David, Mary Gable, and Algernon Austin. 2012. "The Public-Sector Jobs Crisis: Women and African Americans Hit Hardest by Job Losses in State and Local Governments." Economic Policy Institute, May 2. Available at http://www.epi.org/publication/bp339-public-sector-jobs-crisis/.

Cortese, Anthony. 2009. "Higher Education's True Role: Creating a Healthy, Just, and Sustainable Society." In *Green Jobs for a New Economy: The Career Guide to Emerging Opportunities*, edited by Peterson's, 7–9. Lawrenceville, NJ: Peterson's.

Cox, Oliver. 1948. *Caste, Class, and Race: A Study in Social Dynamics*. Garden City, NY: Doubleday.

Currie, Janet. 2011. "Inequality at Birth: Some Causes and Consequences." *American Economic Review: Papers and Proceedings* 101 (3): 1–22.

Davies, Carole Boyce. 2008. *Left of Karl Marx: the Political Life of Black Communist Claudia Jones*. Durham, NC: Duke University Press.

Davis, Angela. 1981. *Women, Race, and Class*. New York: Random House.

Dawson, Michael. 1994. *Behind the Mule: Race and Class in African-American Politics*. Princeton, NJ: Princeton University Press.

———. 2001. *Black Visions: The Roots of Contemporary African-American Political Ideologies*. Chicago: University of Chicago Press.

de Haas, Hein. 2007. "Remittances, Migration and Social Development: A Conceptual Review of the Literature." United Nations Research Institute for Social Development, Social Policy and Development Programme Paper no. 34. Available at http://www.unrisd.org/80256B3C005BCCF9/%28http AuxPages%29/8B7D005E37FFC77EC12573A600439846/$file/deHaas paper.pdf.

Deitche, Scott M. 2010. *Green Collar Jobs: Environmental Careers for the 21st Century*. Santa Barbara, CA: Praeger.

Delmas, Magali A., and Vanessa Cuerel Burbano. 2011. "The Drivers of Greenwashing." *California Management Review* 54 (1): 64–87.

De Sousa, Christopher A. 2005. "Policy Performance and Brownfield Redevelopment in Milwaukee, Wisconsin." *Professional Geographer* 57 (2): 312–327.

De Sousa, Christopher A., Changshan Wu, and Lynne M. Westphal. 2009. "Assessing the Effect of Publicly Assisted Brownfield Redevelopment on Surrounding Property Values." *Economic Development Quarterly* 23:95–110.

Diamond, Jared. 1999. *Guns, Germs, and Steel: The Fates of Human Societies*. New York: W. W. Norton.

Diop, Cheikh Anta. 1974. *The African Origin of Civilization: Myth or Reality*. Translated and edited by Mercer Cook. Westport, CT: Lawrence Hill.

Dobson, Andrew. 2007. *Green Political Thought*. New York: Routledge.

Eagleton, Terry. 2011. *Why Marx Was Right*. New Haven, CT: Yale University Press.

Edsall, Thomas. 2006. *Building Red America: The New Conservative Coalition and the Drive for Permanent Power*. New York: Basic Books.

Eilperin, Juliet. 2013. "On Earth Day, Where Does Obama's Environmental Record Stand?" *Washington Post*, April 22. Available at http://www.wash ingtonpost.com/blogs/the-fix/wp/2013/04/22/on-earth-day-where-does -obamas-environmental-record-stand/.

Eisen, Phyllis, Jerry J. Jasinowski, and Richard Kleinert. 2005. *2005 Skills Gap Report—A Survey of the American Manufacturing Workforce*. Washington, DC: National Association of Manufacturers. Available at http://heartland .org/sites/all/modules/custom/heartland_migration/files/pdfs/18159.pdf.

"Energy Etch a Sketch." 2012. *New York Times*, June 16. Available at http://www.nytimes.com/2012/06/17/opinion/sunday/energy-etch-a-sketch.html.

Etter-Lewis, Gwendolyn. 2006. "The Sky Is Falling: Black Studies and the Politics of Public Discourse." *Journal of Black Studies* 36 (5): 720–731.

Ezzati, Majid. 2005. "Indoor Air Pollution and Health in Developing Countries." *The Lancet* 366:104–106.

Fanon, Frantz. 1963. *The Wretched of the Earth: The Handbook for the Black Revolution That Is Changing the Shape of the World*. New York: Grove Weidenfeld.

Florida, Richard L. 2002. *The Rise of the Creative Class and How It's Transforming Work, Leisure, Community, and Everyday Life*. New York: Basic Books.

Ford, Martin. 2009. *The Lights in the Tunnel: Automation, Accelerating Technology and the Economy of the Future*. Wayne, PA: Acculant.

Foreman, Dave. 1991. *Confessions of an Eco-warrior*. New York: Harmony.

Foster, John Bellamy. 1999. *The Vulnerable Planet: A Short Economic History of the Environment*. New York: Monthly Review Press.

Freudenberg, Nicholas, Manuel Pastor, and Barbara Israel. 2011. "Strengthening Community Capacity to Participate in Making Decisions to Reduce Disproportionate Environment Exposure." *Environmental Justice* 101 (S1): S123–S130.

Fukuyama, Francis. 1992. *The End of History and the Last Man*. New York: Free Press.

General Accounting Office. 1983. *Siting of Hazardous Waste Landfills and Their Correlation with Racial and Economic Status of Surrounding Communities*. Washington, DC: Government Accounting Office. Available at http://www.gao.gov/assets/150/140159.pdf.

Giddings, Paula. 1984. *When and Where I Enter: The Impact of Black Women on Race and Sex in America*. New York: William Morrow.

Gilding, Paul. 2011. *The Great Disruption: Why the Climate Crisis Will Bring On the End of Shopping and the Birth of a New World*. New York: Bloomsbury Press.

Gilmore, Glenda Elizabeth. 2008. *Defying Dixie: The Radical Roots of Civil Rights, 1909–1950*. New York: W. W. Norton.

Glaser, James. 1994. "Back to the Black Belt: Racial Environmental and White Racial Attitudes in the South." *Journal of Politics* 56 (1): 21–44.

Global Humanitarian Forum. 2009. *Human Impact Report: Climate Change*. Geneva: Global Humanitarian Forum. Available at http://www.ghf-ge.org/human-impact-report.pdf.

Goleman, Daniel. 2009. *Ecological Intelligence: How Knowing the Hidden Impacts of What We Buy Can Change Everything*. New York: Broadway Books.

Gonzalez, Priscilla A., Meridith Minker, Analilia P. Garcia, Margaret Gordon, Catalina Garzón, Meena Palaniappan, Swati Prakash, and Brian Beveridge. 2011. "Community-Based Participatory Research and Policy Advocacy to

Reduce Diesel Exposure in West Oakland, California." *American Journal of Public Health* 101 (S1): S166–S175.

Gordon, Vivian G. 1987. *Black Women, Feminism and Black Liberation: Which Way?* Chicago: Third World Press.

Gramsci, Antonio. 1975. *Letters from Prison*. Translated by Lynner Lawner. New York: Harper and Row.

Grey, William. 1993. "Anthropocentrism and Deep Ecology." *Australasian Journal of Philosophy* 71 (4): 463–475.

Grolleau, Gilles, Naoufel Mzoughi, and Sanja Pekovic. 2012. "Greening Not (Only) for Profit: An Empirical Examination of the Effect of Environmental-Related Standards on Employees' Recruitment." *Resource and Energy Economics* 34:74–92.

Guinier, Lani, and Gerald Torres. 2002. *The Miner's Canary: Enlisting Race, Resisting Power, Transforming Democracy*. Cambridge, MA: Harvard University Press.

"Gus Speth: Communicating Environmental Risks in an Age of Disinformation." 2011. *Bulletin of the Atomic Scientists* 67 (4): 1–7.

Guy-Sheftall, Beverly. 1984. *Daughters of Sorrow: Attitudes toward Black Women, 1880–1920*. New York: Carlson.

Hall, Perry A. 2010. "African American Studies: Discourses and Paradigms." In *African American Studies*, edited by Jeanette R. Davidson, 15–34. Edinburg, UK: Edinburg University Press.

Hamilton, James T. 1995. "Testing for Environmental Racism: Prejudice, Profits, Political Power?" *Journal of Policy Analysis and Management* 14 (1): 107–132.

Harlan, Sharon L., and Darren M. Ruddell. 2011. "Climate Change and Health in Cities: Impacts of Heat and Air Pollution and Potential Co-benefits from Mitigation and Adaptation." *Current Opinion in Environmental Sustainability* 3 (3): 126–134.

Harper-Anderson, Elsie. 2012. "Exploring What Greening the Economy Means for African America Workers, Entrepreneurs, and Communities." *Economic Development Quarterly* 26 (2): 162–177.

Hawken, Paul. 2007. *Blessed Unrest: How the Largest Movement in the World Came into Being and Why No One Saw It Coming*. New York: Viking.

Hawken, Paul, Amory Lovins, and L. Hunter Lovins. 1999. *Natural Capitalism: Creating the Next Industrial Revolution*. New York: Back Bay Books.

Hendryx, Michael, and Evan Fedorko. 2011. "The Relationship between Toxics Release Inventory Discharges and Mortality Rates in Rural and Urban Areas of the United States." *Journal of Rural Health* 27 (4): 358–366.

Hertsgaard, Mark. 2011. *Hot: Living through the Next Fifty Years on Earth*. Boston: Houghton Mifflin Harcourt.

Hine, Darlene Clark. 1997. "Black Studies: An Overview." In *Africana Studies: A Disciplinary Quest for Both Theory and Method*, edited by James L. Conyers, Jr., Jefferson, 7–15. Jefferson, NC: McFarland and Company.

hooks, bell. 1982. *Ain't I a Woman: Black Women and Feminism*. Boston: South End Press.

———. 1984. *Feminist Theory: From Margin to Center*. Boston: South End Press.

"Hopeful Continent: Africa Rising." 2011. *The Economist*, December 3. Available at http://www.economist.com/node/21541015.

"Hopeless Africa." 2000. *The Economist*, May 11. Available at http://www.economist.com/node/333429.

Houston, Douglas, Jun Wu, Paul Ong, and Arthur Winter. 2004. "Structural Disparities of Urban Traffic in Southern California: Implications for Vehicle-Related Air Pollution Exposure in Minority and High-Poverty Neighborhoods." *Journal of Urban Affairs* 26 (5): 565–592.

Howland, Marie. 2007. "Employment Effects of Brownfield Redevelopment: What Do We Know from the Literature?" *Journal of Planning Literature* 22:91–107.

Huang, Ganlin, Weiqi Zhou, and M. L. Cadenasso. 2011. "Is Everyone Hot in the City? Spatial Pattern of Land Surface Temperatures, Land Cover and Neighborhood Socioeconomic Characteristics in Baltimore, MD." *Journal of Environmental Management* 92 (7): 1753–1759.

Hudson-Weems, Clenora. 1995. *Africana Womanism: Reclaiming Ourselves*. Troy, MI: Bedford.

Hula, Richard C. 2009. "The Impact of Policy Change in Local and State Environment Policy: Brownfield Redevelopment in Michigan." In *Sustaining Michigan: Metropolitan Policies and Strategies*, edited by Richard Jelier and Gary Sands, 183–206. East Lansing: Michigan State University Press.

International Ecotourism Society. n.d. "What Is Ecotourism?" Available at https://www.ecotourism.org/what-is-ecotourism (accessed August 1, 2014).

Jackson, Tim. 2009. *Prosperity without Growth: Economics for a Finite Planet*. London: Sterling.

James, Winston. 2000. *Holding Aloft the Banner of Ethiopia: Caribbean Radicalism in Early Twentieth-Century America*. London: Verso.

Johnson, Caley, Dylan Hettinger, and Gail Mosey. 2011. "Guide for Identifying and Converting High-Potential Petroleum Brownfield Sites to Alternative Fuel Stations." National Renewable Energy Laboratory, Technical Report NREL/TP-6A20-50898. Available at http://www.afdc.energy.gov/pdfs/50898.pdf.

Johnson, Chalmers. 1982. *MITI and the Japanese Miracle: Growth of Industrial Policy, 1925–1975*. Stanford, CA: Stanford University Press.

Johnson, Daniel P., and Jeffrey S. Wilson. 2009. "The Socio-spatial Dynamics of Extreme Urban Heat Events: The Case of Heat-Related Deaths in Philadelphia." *Applied Geography* 29 (3): 419–434.

Johnson, Simon, and James Kwak. 2010. *13 Bakers: The Wall Street Takeover and the Next Financial Meltdown*. New York: Pantheon Books.

Jones, Van. 2008. *The Green Collar Economy*. New York: HarperOne.

Juhasz, Antonia. 2008. *The Tyranny of Oil: The World's Most Powerful Industry—and What We Must Do to Stop It*. New York: William Morrow.
Karenga. Maulana. 1980. *Kawaida Theory: An Introductory Outline*. Inglewood, CA: Kawaida.
Kelley, Robin D. G. 1994. "Afric's Sons with Banner Red: African American Communists and the Politics of Culture, 1919–1934." In *Imaging Home: Class, Culture and Nationalism in the African Diaspora*, edited by Sidney J. Lemelle and Robin D. G. Kelley, 35–54. New York: Verso.
Kitwana, Bakari. 2002. *The Hip Hop Generation: Young Blacks and the Crisis in African American Culture*. New York: Basic Civitas Books.
———. 2005. *Why White Kids Love Hip Hop: Wankstas, Wiggers, Wannabes, and the New Reality of Race in America*. New York: Basic Civitas Books.
Klein, Ezra. 2011. "What Happened to the 'Fierce Urgency of Now'?" *Washington Post*, April 5. Available at http://www.washingtonpost.com/blogs/wonkblog/post/what-happened-to-the-fierce-urgency-of-now/2011/03/10/AFZOUliC_blog.html.
Kovel, Joel. 2007. *The Enemy of Nature: The End of Capitalism or the End of the World*. London: Zed Books.
Ladner, Joyce. 1971. *Tomorrow's Tomorrow: The Black Woman*. Garden City, NY: Doubleday.
Lange, Deborah, and Sue McNeil. 2004. "Clean It and They Will Come? Defining Successful Brownfield Development." *Journal of Urban Planning and Development* 130 (2): 101–108.
Lee, Charles. 1992. "Toxic Waste and Race in the United States." In *Race and the Incidence of Environmental Hazards: A Time for Discourse*, edited by Bunyan Bryant and Paul Mohai, 10–27. Boulder, CO: Westview Press.
Leftwich, Adrian. 1995. "Bringing Politics Back In: Towards a Model of the Developmental State." *Journal of Development Studies* 31 (3): 400–427.
Legot, Cristina, Bruce London, and John Shandra. 2010. "The Proximity of High Volume Development Neurotoxin Polluters to Schools: Vulnerable Populations at Risk." Political Economy Research Institute, Working Paper Series 224. Available at http://www.peri.umass.edu/fileadmin/pdf/working_papers/working_papers_201-250/WP224.pdf.
Lenin, Vladimir. 1977. *Collected Works*, vol. 29. Moscow: Progress.
Leopold, Aldo. 1949. *A Sand Country Alamanac, and Sketches Here and There*. New York: Oxford University Press.
Lerner, Gerda. 1973. *Black Women in White America*. New York: Vintage Books.
Lester, James P., David W. Allen, and Kelly M. Hill. 2001. *Environmental Injustice in the United States: Myths and Realities*. Boulder, CO: Westview Press.
Lisell, Lars, and Gail Mosey. 2010. "Feasibility Study of Economics and Performance of Solar Photovoltaics at the Former St. Marks Refinery in

St. Marks, Florida." National Renewable Energy Laboratory, Technical Report NREL/TP-6A2-48853. Available at http://www.nrel.gov/docs/fy10osti/48853.pdf.

Liu, Yingling. 2008. "China's New Path." *Our Planet*, December, pp. 10–11. Available at http://www.unep.org/PDF/ourplanet/2008/dec/en/OP-2008-12-en-ARTICLE3.pdf.

Lodge, Tom. 2009. "The South African Developmental State?" *Journal of Southern African Studies* 35 (1): 253–261.

Lorde, Audre. 1984. "Age, Race, Class, and Sex: Women Redefining Difference." In *Sister Outsider: Essays and Speeches*, 114–123. Freedom, CA: Crossing Press.

Lucier, Cristina, Anna Rosofsky, and Bruce London. 2011. "Toxic Pollution and School Performance Scores: Environmental Ascription in East Baton Rouge Parish, Louisiana." *Organization and Environment* 24:423–443.

Marable, Manning. 1995. *Beyond Black and White: Transforming African-American Politics*. London: Verso.

Marina, Daniel Perez. 2009. "Anthropocentrism and Androcentrism: An Ecofeminist Connection." Unpublished manuscript. School of Culture and Communication, Södertörn University College, Flemingsberg, Sweden.

McCarthy, Linda. 2009. "Off the Mark? Efficiency in Targeting the Most Marketable Site rather than Equity in Public Assistance for Brownfield Redevelopment." *Economic Development Quarterly* 23:211–228.

McKibben, Bill. 1989. *The End of Nature*. New York: Random House.

———. 2010. *Eaarth: Making a Life on a Tough New Planet*. New York: Times Books.

McKinsey Global Institute. 2010. "Lions on the Move: The Progress and Potential of African Economies." Available at http://www.mckinsey.com/insights/africa/lions_on_the_move.

Mies, Maria. 1986. *Patriarchy and Accumulation on a World Scale: Women in the International Division of Labour*. London: Zed.

Miller, G. Tyler. 1972. *Replenish the Earth: A Primer on Human Ecology*. Belmont, CA: Wadsworth.

Miranda, Marie Lynn, Martha H. Keating, and Sharon E. Edwards. 2008. "Environmental Justice Implications of Reduced Reporting Requirements of the Toxics Release Inventory Burden Reduction Rule." *Environmental Science and Technology* 42 (15): 5407–5414.

Mohai, Paul, and Robin Saha. 2006. "Reassessing Racial and Socioeconomic Disparities in Environmental Justice Research." *Demography* 43 (2): 383–399.

Mol, Arthur P. J. 1996. "Ecological Modernisation and Institutional Reflexivity: Environmental Reform in the Late Modern Age." *Environmental Politics* 5 (2): 302–323.

Morrone, Michele, and Geoffrey L. Buckley, eds. 2011. *Mountains of Injustice: Social and Environmental Justice in Appalachia*. Athens: Ohio University Press.

Muir, John. 1913. *The Story of My Boyhood and Youth*. Boston: Houghton Mifflin. Available at http://www.yosemite.ca.us/john_muir_writings/the_story_of_my_boyhood_and_youth/chapter_8.html.

Murphy, David J., and Charles A. S. Hall. 2011. "Energy Return on Investment, Peak Oil, and the End of Economic Growth," in "Ecological Economics Reviews," edited by Robert Costanza, Karin Limburg, and Ida Kubiszewski. Special issue, *Annals of the New York Academy of Sciences* 1219:52–72.

Naess, Arne. 1973. "The Shallow and the Deep, Long-Range Ecology Movement." *Inquiry* 16:95–100.

Nagengast, Amy, Chris Hendrickson, and Deborah Lange. 2011. "Commuting from U.S. Brownfield and Greenfield Residential Development Neighborhoods." *Journal of Urban Planning and Development* 137 (3): 298–304.

New Partnership for Africa's Development. 2003. *Action Plan for the Environment Initiative*. Johannesburg: NEPAD.

Norment, Nathaniel, Jr., ed. 2007. *The African American Studies Reader*. Durham, NC: Carolina Academic Press.

Norton, Bryan. 1984. "Environmental Ethics and Weak Anthropocentrism." *Environmental Ethics* 6 (2): 131–148.

O'Connor, James. 1998. *Natural Causes: Essays in Ecological Marxism*. New York: Guilford Press.

Okri, Ben. 2010. "The Moral Bankruptcy of Our Civilization." In *The Spiritual Life of Water: Its Power and Purpose*, by Alick Bartholomew, 310–312. Rochester, VT: Park Street Press.

OMB Watch. 2006. "Against the Public's Will: Summary of Responses to the Environmental Protection Agency's Plans to Cut Toxic Reporting." Available at http://www.ombwatch.org/files/info/TRICommentsReport.pdf.

O'Rourke, Dara, and Gregg P. Macey. 2003. "Community Environmental Policing: Assessing New Strategies of Public Participation in Environmental Regulation." *Journal of Policy Analysis and Management* 22 (3): 383–414.

Pastor, Manuel, Jr., Jim Sadd, and John Hipp. 2001. "Which Came First? Toxic Facilities, Minority Move-in, and Environmental Justice." *Journal of Urban Affairs* 23 (1): 1–21.

Patterson, Rubin. 2006. "Transnationalism: Diaspora-Homeland Development." *Social Forces* 84:1891–1907.

———. 2007. "Going around the Drain-Gain Debate with Brain Circulation." In *African Brain Circulation: Beyond the Drain-Gain Debate*, edited by Rubin Patterson, 1–14. Boston: Brill.

———. 2008. "Preparing Sub-Saharan Africa for a Pioneering Role in Eco-industrial Development." *Journal of Industrial Ecology* 12:501–504.

———. 2010. "A Great Dilemma Generates Another Great Transformation: Incompatibility of Capitalism and Sustainable Environments." *Perspectives on Global Development and Technology* 9:74–83.

———. 2011. "Renewable Energy, Migration-Development Model, and Sustainability Entrepreneurship." In *Globalization and Sustainable Development in Africa*, edited by Bessie House-Soremekun and Toyin Falola, 103–124. Rochester, NY: University of Rochester Press.

———. 2012. "Growing the Global Green Economy: Getting Africa Prepared to Lend a Hand." In *Landscape, Environment and Technology in Colonial and Postcolonial Africa*, edited by Toyin Falola and Emily Brownell, 311–327. New York: Routledge.

———. 2013a. "Historic Changes Underway in African Migration Policies: From Muddling through to Organized Brain Circulation." In *African Patterns of Migration in a Global Era: New Perspectives*, edited by Abdoulaye Kane and Todd Leedy, 78–89. Bloomington: Indiana University Press.

———. 2013b. "The Transnational Capitalist Class: What's Race Got to Do with It? Everything!" *Globalizations* 10 (5): 674–690.

Patterson, Rubin, and Ernest J. Wilson. 2000. "IT and Social Inequality: Resetting the Research and Policy Agendas." *Information Society: An International Journal* 16 (1): 77–86.

Peery, Nelson. 2007. *Black Radical: The Education of an American Revolutionary*. New York: New Press.

Pellow, David Naguib. 2007. *Resisting Global Toxics: Transnational Movements for Environmental Justice*. Cambridge, MA: MIT Press.

Perlin, Susan A., David Wong, and Ken Sexton. 2001. "Residential Proximity to Industrial Sources of Air Pollution: Interrelationships among Race, Poverty, and Age." *Journal of the Air and Waste Management Association* 51 (3): 406–421.

Phillips, Kevin. 2006. *American Theocracy: The Peril and Politics of Radical Religion, Oil, and Borrowed Money in the 21st Century*. New York: Penguin.

Plumwood, Val. 1997. "Androcentrism and Anthropocentrism: Parallels in Politics." In *Ecofeminism: Women, Culture, Nature*, edited by Karen Warren, 327–355. Bloomington: Indiana University Press.

Pojman, Louis P., and Paul Pojman. 2012. *Environmental Ethics: Reading in Theory and Application*. 6th ed. Boston: Wadsworth.

Polgreen, Lydia. 2012. "U.S., Too, Wants to Bolster Investment in a Continent's Economic Promise." *New York Times*, August 8. Available at http://www.nytimes.com/2012/08/09/world/africa/us-seeks-to-step-up-africa-investment.html?pagewanted=all&_r=0.

Pollin, Robert, Mark Brenner, Jeanette Wicks-Lim, and Stephanie Luce. 2008. *A Measure of Fairness: The Economics of Living Wages and Minimum Wages in the United States*. Ithaca, NY: Cornell University Press.

Rabaka, Reiland. 2009. *Africana Critical Theory: Reconstructing the Black Radical Tradition, from W.E.B. Du Bois and C.L.R. James to Frantz Fanon and Amilcar Cabral*. New York: Lexington Books.

Radelet, Steven. 2010. *Emerging Africa: How 17 Countries Are Leading the Way*. Washington, DC: Center for Global Development.

Rattner, Steven. 2012. "The Rich Get Even Richer." *New York Times*, March 26, p. A23.

Rhodes, Edwardo Lao. 2003. *Environmental Justice in America*. Bloomington: Indiana University Press.

Rifkin, Jeremy. 2011. *The Third Industrial Revolution: How Lateral Power Is Transforming Energy, the Economy, and the World*. New York: Palgrave Macmillan.

Robinson, Cedric J. 2000. *Black Marxism: The Making of the Black Radical Tradition*. Chapel Hill: University of North Carolina Press.

Robinson, Mark, and Gordon White. 1998. *The Democratic Developmental State: Politics and Institutional Design*. New York: Oxford University Press.

Rojas, Fabio. 2007. *From Black Power to Black Studies: How a Radical Social Movement Became an Academic Discipline*. Baltimore: Johns Hopkins University Press.

Rom, William N. 2011. *Environmental Policy and Public Health: Air Pollution, Global Climate Change, and Wilderness*. Hoboken, NJ: Jossey-Bass.

Roosevelt, Margot. 2010. "Prop. 23: Why Did Valero Launch a Campaign against California's Climate Law?" *Los Angeles Times*, October 31. Available at http://latimesblogs.latimes.com/greenspace/2010/10/prop-23-valero-global-warming-oil-refineries.html.

Sandler, Ronald, and Phaedra Pezzullo, eds. 2007. *Environmental Justice and Environmentalism: The Social Justice Challenge to the Environmental Movement*. Cambridge, MA: MIT Press.

Saxenian, AnnaLee. 2006. *The New Argonauts: Regional Advantage in a Global Economy*. Cambridge, MA: Harvard University Press.

Schweickart, David. 2011. *After Capitalism*. Lanham, MD: Rowman and Littlefield.

Semmes, Clovis E. 1992. *The Cultural Hegemony and African American Development*. Westport, CT: Praeger.

Shelby, Tommie. 2005. *We Who Are Dark: The Philosophical Foundations of Black Solidarity*. Cambridge, MA: Harvard University Press.

Shi, Zhengrong. 2009. "Solar Solution." *Our Planet*, February, pp. 6–7. Available at http://www.unep.org/PDF/ourplanet/2009/feb/en/OP-2009-02-en-ARTICLE2.pdf.

Shiva, Vandana. 1989. "Development, Ecology, and Women." In *Healing the Wounds: The Promise of Ecofeminism*, edited by Judith Plant, 80–90. London: Zed Books.

———. 2010. *Staying Alive: Women, Ecology and Development*. Cambridge, MA: South End Press.

Smith, Zachary A. 2012. *The Environmental Policy Paradox*. 6th ed. New York: Pearson.

So, Alvin Y. 1990. *Social Change and Development: Modernization, Dependency, and World-System Theories*. Newbury Park, CA: Sage.

Speth, James Gustave. 2008. *The Bridge at the Edge of the World: Capitalism, the Environment, and Crossing from Crisis to Sustainability*. New Haven, CT: Yale University Press.

Stephens-Davidowitz, Seth. 2012. "How Racist Are We? Ask Google." *New York Times*, June 10, p. 12.

Stewart, James B. 1992. "Reaching for Higher Ground: Toward an Understanding of Black/Africana Studies." *Afrocentric Scholar* 1 (1): 1–63.

Stoecker, Randy. 1997. "The CDC Model of Urban Redevelopment: A Critique and an Alternative." *Journal of Urban Affairs* 19 (1): 1–22.

Strietska-Ilina, Olga, Christine Hofmann, Mercedes Durán Haro, and Shinyoung Jeon. 2011. *Skills for Green Jobs: A Global View*. Geneva: International Labour Office.

Taylor, Dorceta E. 2000. "The Rise of the Environmental Justice Paradigm: Injustice Framing and the Social Construction of Environment Discourses." *American Behavioral Scientist* 43:508–580.

———. 2002. "Race, Class, Gender, and American Environmentalism." U.S. Department of Agriculture General Technical Report PNW-GTR-534. Available at http://wcsu.csu.edu/cerc/documents/RaceClassGenderandAmericanEnvironmentalism.pdf.

———. 2005. "American Environmentalism: The Role of Race, Class and Gender in Shaping Activism, 1820–1995." In *Environmental Sociology: From Analysis to Activism*, edited by Leslie King and Deborah McCarthy, 87–106. Lanham, MD: Rowman and Littlefield.

———. 2007a. "Diversity and Equity in Environmental Organizations: The Salience of These Factors to Students." *Journal of Environmental Education* 39 (1): 19–43.

———. 2007b. "Employment Preferences and Salary Expectations of Students in Science and Engineering." *BioScience* 57 (2): 175–185.

———. 2008. "Diversity and the Environment: Myth-Making and the Status of Minorities in the Field. *Equity and the Environment* 15:89–147.

Tiefenbacher, John P., and Ronald R. Hagelman, III. 1999. "Environmental Equity in Urban Texas: Race, Income, and Patterns of Acute and Chronic Toxic Air Releases in Metropolitan Counties." *Urban Geography* 19:516–533.

United Nations. 1987. "Report of the World Commission on Environment and Development: Our Common Future." Available at http://conspect.nl/pdf/Our_Common_Future-Brundtland_Report_1987.pdf.

United Nations Development Programme. 2007. *Human Development Report 2007/2008: Fighting Climate Change; Human Solidarity in a Divided World*. Basingstoke, UK: Palgrave Macmillan. Available at http://hdr.undp.org/sites/default/files/reports/268/hdr_20072008_en_complete.pdf.

———. 2011. *Human Development Report 2011: Sustainability and Equity; A Better Future for All*. New York: UNDP. Available at http://www.undp.org/content/dam/undp/library/corporate/HDR/2011%20Global%20HDR/English/HDR_2011_EN_Complete.pdf.

———. 2012. *Africa Human Development Report 2012: Towards a Food Secure Future*. New York: UNDP. Available at http://www.undp.org/content/dam/

undp/library/corporate/HDR/Africa%20HDR/UNDP-Africa%20HDR-2012-EN.pdf.
———. 2013. *Human Development Report 2013; The Rise of the South*. New York: UNDP.
United Nations Environment Programme. n.d. "Cities and Buildings." Available at http://www.unep.org/SBCI/pdfs/Cities_and_Buildings-UNEP_DTIE_Initiatives_and_projects_hd.pdf (accessed August 4, 2014).
———. 2005. *AEO for Youth: Africa Environment Outlook for Youth; Our Region—Our Life*. Nairobi, Kenya: UNEP. Available at http://www.unep.org/geo/pdfs/AEO_for_youth_report.pdf.
———. 2008. *Green Jobs: Towards Decent Work in a Sustainable, Low-Carbon World*. Nairobi, Kenya: UNEP. Available at http://www.unep.org/PDF/UNEPGreenjobs_report08.pdf.
———. 2010. "Driving a Green Economy through Public Finance and Fiscal Policy Reform." Available at http://www.unep.org/greeneconomy/Portals/88/documents/ger/GER_Working_Paper_Public_Finance.pdf.
U.S. Conference of Mayors. 2008. "U.S. Metro Economies: Current and Potential Green Jobs in the U.S. Economy." Available at http://www.usmayors.org/pressreleases/uploads/greenjobsreport.pdf.
U.S. Department of Energy. 2002. "U.S. Department of Energy Weatherization Assistance Program." Available at http://www.nrel.gov/docs/fy02osti/31147.pdf.
U.S. Department of Labor. 2012. "Green Jobs for Women." Available at http://www.dol.gov/wb/media/green.htm.
U.S. Energy Information Administration. 2012a. "Annual Energy Review." Available at http://www.eia.gov/totalenergy/data/annual/showtext.cfm?t=ptb1001.
———. 2012b. *Monthly Energy Review, March 2012*. Washington, DC: U.S. Energy Information Administration. Available at http://www.eia.gov/totalenergy/data/monthly/archive/00351203.pdf.
———. 2013. "What Are the Major Sources and Users of Energy in the United States?" Available at http://www.eia.gov/energy_in_brief/major_energy_sources_and_users.cfm.
U.S. Environmental Protection Agency. 2010. "Fiscal Year 2011–2015 EPA Strategic Plan." Available at http://nepis.epa.gov/Exe/ZyPDF.cgi?Dockey=P1008YOS.PDF.
———. 2012. "Brownfields and Land Revitalization: Basic Information." Available at http://www.epa.gov/brownfields/basic_info.htm.
U.S. Office of Personnel Management. 2006. "Table 2—Race/National Origin Distribution of Federal Civilian Employment by Payplan and Grade as of September 30, 2006." Available at http://www.opm.gov/feddata/demograp/Table2mw.pdf.
Valentino, Nicholas A., and David O. Sears. 2005. "Old Times There Are Not Forgotten: Race and Partisan Realignment in the Contemporary South." *American Journal of Political Science* 49 (3): 672–688.

Walekhwa, Peter N., Johnny Mugisha, and Lars Drake. 2009. "Biogas Energy from Family-Sized Digesters in Uganda: Critical Factors and Policy Implications." *Energy Policy* 37 (7): 2754–2762.

Wapner, Paul. 2010. *Living through the End of Nature: The Future of American Environmentalism*. Cambridge, MA: MIT Press.

Warren, Karen J. 1993. "Ecological Feminist Philosophies: An Overview of the Issues." In *Ecological Feminist Philosophies*, edited by Karen J. Warren, ix–xxvi. Available at http://www.vedegylet.hu/okopolitika/Warren%20-%20Ecofeminism%20Overview.pdf.

———. 1997. "Taking Empirical Data Seriously: An Ecofeminist Philosophical Perspective." In *Ecofeminism: Women, Culture, Nature*, edited by Karen Warren, 3–20. Bloomington: Indiana University Press.

West, Cassandra. 2012. "Black Studies Programs Now Flourishing despite Early Struggles." *Diverse Issues in Higher Education* 29:10–12.

West, Cornel. 1988. "Marxist Theory and the Specificity of Afro-American Oppression." In *Marxism and the Interpretation of Culture*, edited by Cary Nelson and Lawrence Grossberg, 17–33. Urbana: University of Illinois Press.

Wilson, William Julius. 1996. *When Work Disappears: The World of the New Urban Poor*. New York: Vintage Books.

Winant, Howard. 2000. "Race and Race Theory." *Annual Review of Sociology* 26:169–185.

World Travel and Tourism Council. 2012. "Progress and Priorities, 2011–2012." Available at http://92.52.122.233/site_media/uploads/downloads/Progress_Priorities_2011_2012.pdf.

Zhu, Qinghua, Joseph Sarkis, and Kee-hung Lai. 2012. "Green Supply Chain Management Innovation Diffusion and Its Relationship to Organizational Improvement: An Ecological Modernization Perspective." *Journal of Engineering and Technology Management* 29 (1): 168–185.

Index

Rubin Patterson is Professor and Chair of Sociology and Anthropology at Howard University in Washington, D.C. He is also a Research Associate in the Department of Sociology at the University of the Witwatersrand in Johannesburg, South Africa. He has published essays on transnationalism and environmental issues in various journals, including *Social Forces*, *Globalizations*, the *Journal of Black Studies*, and the *Journal of Industrial Ecology*. He served for ten years as the founding editor of the journal *Perspectives on Global Development and Technology*.